食物小传

糖

Sugar

A Global History

〔美国〕安德鲁·F. 史密斯 著
王艺蒨 译

北京联合出版公司
Beijing United Publishing Co.,Ltd.

目录

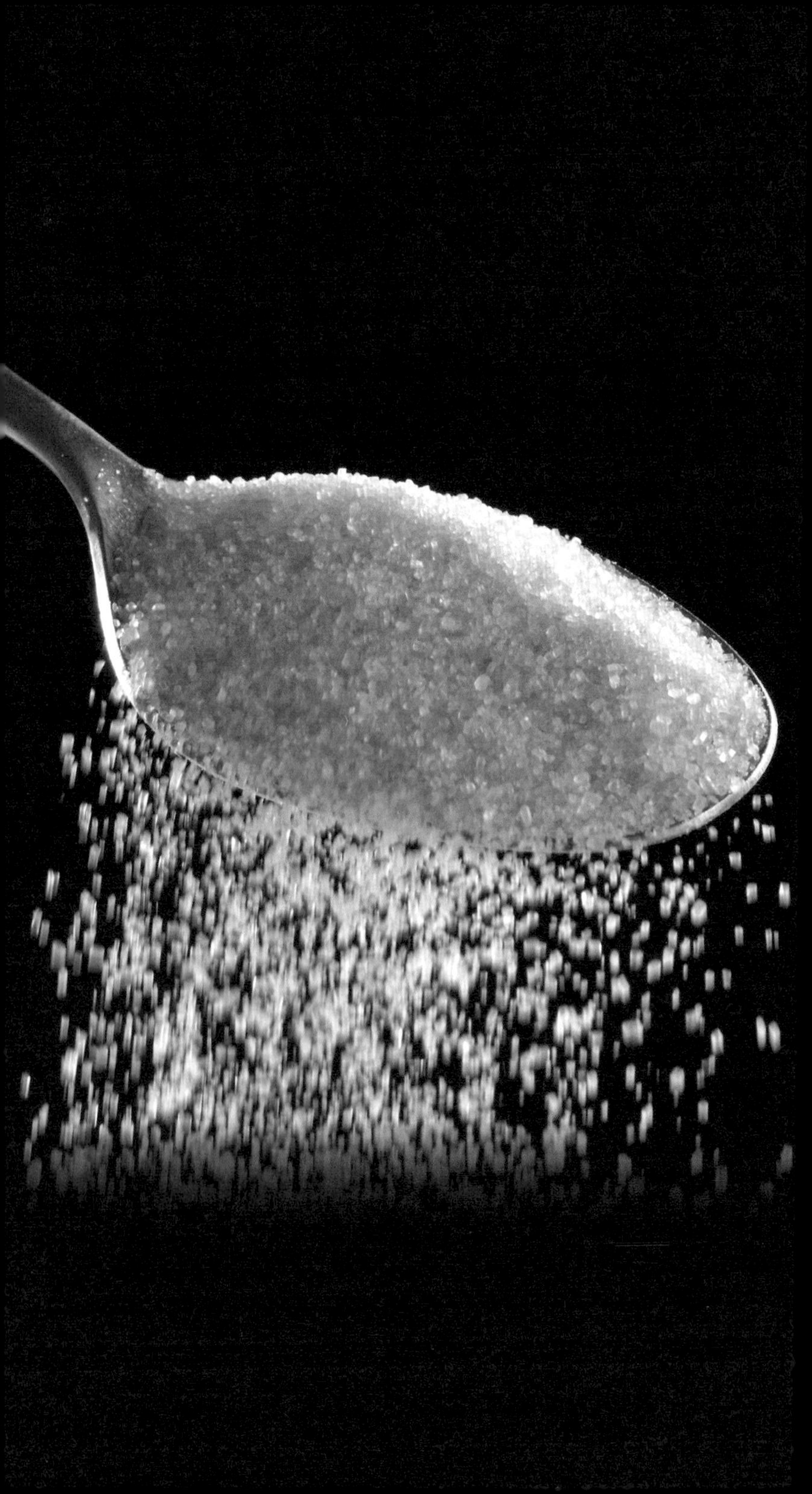

序言

自出生之日起，人类就会受到甜味的吸引，当然，这是有合理原因的：口腔中的上万个味蕾全都有专门的甜味受体。甜的食物可以刺激味蕾分泌神经递质，激活人脑的愉悦中枢。而大脑因此产生内源性大麻素，进一步增进我们的食欲。从进化学的角度来看，母乳中大概 40% 的热量都是由乳糖提供的，这是一种双糖（又名二糖），能很快被代谢成葡萄糖，是人体的基本燃料。这种甜味让婴儿多多进食，从而更有可能活下来。

天然的苦味植物往往是在警示自身的毒性，而甜的食物通常可以被安全食用，而且是不错的简单碳水化合物来源。一旦我们习惯了吃甜食，光是看到它们就足以让我们垂涎三尺。唾液会帮助人体开启分解碳水化合物的过程，向消化系统发出信号，告诉身体“营养物质已经在来的路上了”。

几千年来，我们的先祖种植和培育了甜的蔬果，还会用水果、莓果、无花果、枣、坚果、胡萝卜的汁水，角豆树、枫树或者棕榈树的树液，花蜜和香草的叶片与种子，给食物增加甜味。在过去的几个世纪里，人类学会了收获、提炼或浓缩甜味剂，比如从谷物中获取麦芽糖，从葡萄中得到葡萄糖，从水果、莓果和玉米里提取果糖，还有从甘蔗和甜菜里榨取蔗糖。人们甚至能利用蜜蜂来获取蜂蜜，这正是旧大陆第一种重要的甜味剂。

然而，在过去五百年间最常见的甜味剂实际上是蔗糖（$C_{12}H_{22}O_{11}$）——一种由葡萄糖和果糖两种单糖通过化学组合连接而成的双糖。在消化的过程中，两者又会分开；葡萄糖分子通过小肠流进血液，进而输送到各个器官，被代谢为能量（任何多余的、不需要的能量都会储存在脂肪细胞中）。而果糖这种甜度最高的天然甜味剂主要是在肝脏中被代谢，酶会把它转化为葡萄糖。

大多数植物都含有蔗糖，但人们发现含量最高的是甘蔗属——一类很高大、长得像竹子一样的禾本科植物。这个属很可能发源于南亚或东南亚，包含几个物种，而每个物种又都有各自的变种。只有新几内亚野生蔗和细茎野生种两种甘蔗可以在野外繁殖，而且含糖量都相对较低。新几内亚野生蔗原产自新几内亚，当地人又培育出了比其他品种含糖量更高的秀贵甘蔗，也称克里奥尔甘蔗。这项壮举极为成功，使得秀贵甘蔗在 8000 年前就广泛传播，踪迹遍及菲律宾、印度尼西亚、印度、东南亚和中国。在印度，秀贵甘蔗和原产于南亚的甘蔗细茎野生种杂交后产生了细秆甘蔗，并在印度得到普及。而中国也同样将秀贵甘蔗和细茎野生种杂交，这次得到了中国竹蔗，是中国南部常见的甘蔗品种。

几千年来，人们种植和开发了各种饱含甜味汁液的甘

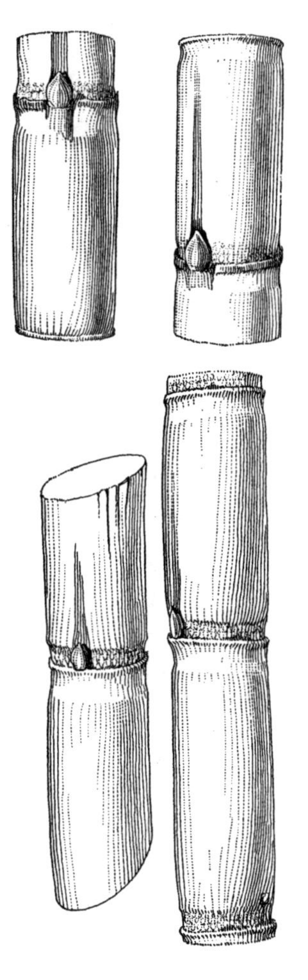

甘蔗的“眼”或者“结”

蔗种类，虽然从18世纪末以来其他的品类和变种会被用来杂交育种，但热带种甘蔗一直牢牢占据着蔗糖产业的核心地位。种植和加工甘蔗是一项劳动强度高的活动。所有驯化的甘蔗都是无性繁殖的——带有至少一个芽的甘蔗茎秆

（也称“眼”或者“结”）会被截成一段，种到田里。许多甘蔗田里需要除草、施肥和灌溉。成熟之后，甘蔗会被整根砍下来。在20世纪机械工具被发明出来之前，这些劳作全部都是靠人们的双手完成的。

在理想的环境下，甘蔗秆在生长期的几周内可以每天生长5厘米（约2英寸）。到了成熟的时候，它们会长到约5厘米粗，3.6~4.6米高（约12~15英尺）。到9~18个月时，它们的任意部分的含糖量都能达到最佳状态。当甘蔗茎开花的时候，蔗糖含量就达到了最高水平（理想的话是17%）。甘蔗秆会被从根部上方一点的地方砍下来，这个过程叫作“留根耕种”。根系会再生出新的茎秆，但含糖量会低一些，抗病能力也会变差。不过这株甘蔗还需要再留根耕种几次才会被彻底铲除，这时再种一棵新甘蔗才会更划算一些。

人类对于种植甘蔗心甘情愿的付出，清楚地体现了几千年来我们对于甜味的青睐。最开始，人们品尝甘蔗的方式就是简单咀嚼或者吮吸甘蔗的茎片。将砍下来的甘蔗保存或者贮藏稍微长一点时间都是很困难的事：一旦砍下，茎秆很快就会腐败，变成棕色的烂泥。把甘蔗挤出汁也是一个办法，但汁水一旦接触空气就会开始发酵。如果期望得到的最终产品是酒的话，这一特性无疑是个优势；但如果是想获得可以保存的甜味剂，就不起作用了。本书要讲述的正是，我们的祖先是如何找到储存甘蔗汁的办法，以及这一工艺的实施和改进是如何影响人类历史的。

［第一章］

糖的早期历史

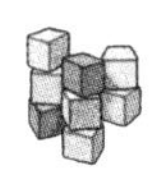

从甘蔗中榨取甜味的汁液，再把它变成结晶的糖，是一个复杂的过程。几乎没有考古证据可以显示甘蔗汁是具体在何时何地被第一次转化为可供长期保存的糖的。很多历史学家认为，大约2500年前的东印度是蔗糖产业的发源地。这个说法主要缘于很多早期的印度文字资料里面都提到了蔗糖和它的甜汁水。波颠阇利[①]在公元前400—前200年之间所著的注释梵文文法的《大疏》里就收录了米布丁、大麦粉和发酵饮料的食谱——这些食物都用了糖类来增甜。

大致在公元前324—前300年，考底利耶[②]讨论权谋治国之道的经典梵文著作《政事论》也提到了糖。这本书描述了不同的糖制品，比如“guda”[③]（最粗制的一类糖）到“khanda”[④]（英语中“candy”一词的来源）和“śarkarā”[⑤]，也就是最纯净的糖。在古代，“śarkarā”大概类似于直到今天还在使用的印度甜味剂，现在叫椰糖——一种粗糙的固体糖，含有一些糖蜜、灰烬和其他杂质。（讽刺的是，“śarkarā”这个梵语词汇到了英文中最终竟然变成了“糖精”——一种代糖。）

早期制糖，要用牲畜拉装有石轮的磨盘，把甘蔗秆压碎研磨，和那时磨谷物的方法差不多。通过碾压，汁水

① Patanjali，古印度哲学家。——本书注释，除特别说明，均为编注
② Kautilya，古印度政治家、哲学家。
③ 梵文，指粗制糖。
④ 梵文，指一种质地纯净、颜色白亮的糖。
⑤ 梵文，指一种比 khanda 更白更纯的糖，它的形状宛如沙粒。

被挤出来，再经过熬煮使其浓缩。剩下的就是原糖了，这种呈脏兮兮棕色的半固体糖不会发酵变质。随着时间的流逝，人们发明出一些过滤杂质的方法，得到了更白净、更甜、更透明的成品。结晶可以从深色的液体中被分离出来，然后塑形成软质的球体。之后，它们会被制成固态片状的硬糖；最终，在需要时被磨成砂糖颗粒。那些粗糙、深色的液体后来被称为糖蜜，会在研磨这一程序中被舍弃掉。那时的技术无法将这种液体进一步提纯出糖，但它可以被用作甜味剂来制作酒精类饮料。

精制糖的好处多不胜数。它可以被磨成颗粒或者粉末、晶化、融化、拉丝、拉扯、煮沸和塑形。无论是在家庭厨房中还是工业产品流水线上，它都能和其他的原材料完美地融合。它可以被用来掩盖苦涩，增强药物的功效。它还能保存很长一段时间，让我们的甜味终年不断。精制糖在烹饪中更是大有用处，比如削弱或是增强一些味道，制作酒精饮料以及储存水果和蔬菜。同时，精制糖可以被运到甘蔗无法生长的地方，成为早期重要的贸易商品。

广泛种植、加工甘蔗的东印度也是佛教的诞生地。《中国：糖与社会》（1998）的作者穆素洁指出，甘蔗被引入很多佛教仪式中，而且释迦牟尼佛祖（公元前 563—前 483）很多的名言都提及了甘蔗。斋戒的僧侣可以喝含糖的果汁，而且很多佛教节庆食物也是用糖制作而成的。

糖的身影在其他一些早期宗教里也不难寻觅，例如觉音大师的印度教著作，以及描写了蔗糖工厂、甘蔗汁，煮沸甘蔗汁以及原糖和糖块的《论道德意识》（约公元 500 年）。有些研究糖的历史学家认为，书中所写的糖块是有延展性的而不是硬的，就像太妃糖一样；也有人认为这就是结晶糖在历史舞台的首次亮相。耆那教文献中也提到了一种糖，它对不吃蜂蜜的耆那教徒非常重要。因为他们认为蜂蜜中孕育着数以万计的生命体，吃蜂蜜会把这些生命杀死。

我们暂且认为印度是制糖的发源地，这项技术很快就传到了东南亚和中国南方。除了知道糖——很可能是以雕塑的形式——在公元前 221 年出口到了中国之外，人们对东南亚的经营业务知之甚少。而关于中国早期制糖业的信息则更多一些，据说是佛教徒把甘蔗和制作固体糖的方法引入此地的。即使僧侣不是最初的引进者，也一定是他们让糖在中国风靡起来的。

然而，蔗糖或者甘蔗浆并不是中国最早的甜味剂。在以谷物为主要作物的中国北方，人们以高粱为主料制作出了一种成分为麦芽糖的糖蜜。麦芽糖是一种由两个葡萄糖分子组成的双糖，比蔗糖的甜度低不少。它是中国最为重要的甜味剂，直到今天还会用在中式烹饪中。

在公元前 3 世纪，制作蔗糖的方法被引入了中国南方，但这种工艺直到几个世纪以后才在北方地区普及。中国人通常会用糖制药，也会用糖为食物和饮品增加甜度；中国人很有可能是第一批造出冰糖的人。然而，蔗糖并非中国人的必需品，而且中国制糖工艺也没有像东南亚和后来的中东地区那样发展。据 13 世纪末游历中国的马可 · 波罗所说，中国的元朝皇帝忽必烈引进了埃及的制糖匠人来教中国人如何生产蔗糖。中东地区的甘蔗种植、蔗糖制造和准备工艺确实取得了重大的进步。

中东和地中海地区的糖

古希腊人和古罗马人曾到访印度，并且注意到了印度的糖。亚历山大大帝的将军尼阿库斯在公元前 327 年乘船从印度河河口驶向小亚细亚的幼发拉底河河口，在他的《印度记》中写道："印度有种芦苇，无须蜂的帮助就能酿出蜜，无须结果就能制成让人欲罢不能的饮品。"

在古罗马时期，少量的糖进入了地中海地区。这些进口货物是用来制药的。公元 1 世纪的希腊医生与植物学家迪奥斯科里季斯在他的五卷本《药理》中写道："有一种在印度和阿拉伯菲利克斯[①]的芦苇中发现的凝固状蜂蜜，叫作'糖精'。"他又补充道："它形似盐巴，像盐一样脆。"盖伦、塞涅卡、普林尼和其他一些作家也都提到了一种从印度进口的甜味剂，现代学者认为这些指的应该就是糖。到了公元 6 世纪，糖从印度被船运到索马里沿岸的港口，接着通过陆路转运到亚历山大港，又从那里以很少的分量卖给医师来满足医疗需求。

公元 600 年前后，甘蔗已经在美索不达米亚的土地上

① Arabia Felix，希腊语名，意为幸福繁荣的阿拉伯，过去的地理学家用来形容阿拉伯南部地区或者现在的也门。

生根发芽了，不久之后就开始了商业化的生产。根据拜占庭历史学家塞奥法尼斯的记载，糖块是公元 622 年赫拉克利乌斯击溃萨珊王朝之后缴获的价值极高的战利品之一。阿拉伯人在 641 年征服了美索不达米亚，经他们之手，甘蔗和制糖工艺向西传播，传到了尼罗河、尼罗河三角洲、地中海东部和东非地区。相关的作物和技法的传播并未止步于此，它们继续西进，来到了地中海岛屿——塞浦路斯、马耳他、克里特、西西里和罗德岛——甘蔗在北非被广泛种植，于公元 682 年到达了摩洛哥南部。在这之后，甘蔗还陆续在西班牙南部的一些地方、意大利南部和土耳其被种植。

在美索不达米亚，制糖工业的中心就在波斯湾的源头，也就是底格里斯－幼发拉底河三角洲。糖成为巴格达的主要商品，巴格达当时掌握着一个横跨如今的伊朗和西班牙的帝国。据估计，巴格达有大概一百万的人口，是当时世界上最大的城市。伊本·萨亚尔·瓦拉克在 10 世纪撰写的巴格达烹饪书中就有许多以糖为原料的食谱。在蒙古人大举进攻并于 1258 年洗劫巴格达之前，这里制糖工业的发展都很兴盛。后来这个地区陷入了政治混乱，制糖生产遭到破坏，好在这时地中海区域的制糖工业已经发展成熟。

在中东和地中海地区种植甘蔗和制作蔗糖需要比在印度投入更高额的资金。印度东部因为干燥炎热的气候需要庞大的灌溉系统才能种植甘蔗。而这些设施常常要延伸很长一段距离，因此需要政府或者阔绰的地主来建造、维护和管理灌溉系统。同样，愿意买并且买得起甘蔗的顾客也是必不可少的——毕竟这可是当时昂贵的奢侈品。

上埃及的地理位置很适合种植甘蔗。仰赖温暖的气候以及尼罗河三角洲丰沛的水资源和肥沃的土壤，糖成了埃及人饮食生活中的重要食材，特别是在有钱人家的厨房里。不过有的时候，糖也会分发给平民百姓。宴席上通常会放置糖雕，依据宾客不同的等级，他们能分到 1~25 磅（约 0.45~11.3 千克）不等的糖作为礼物。糖还会在上埃及的市场上出售,那里是中东地区和欧洲主要的糖类货源地。蔗农、蔗糖磨坊主和提炼商因此变得富裕起来。

当欧洲人重新夺回像克里特和西西里这样的地中海领土之后，他们也从阿拉伯人那里学会了如何种植甘蔗和制糖。十字军东征期间，欧洲人征服了耶路撒冷并开启了从 1099 年到 1187 年的统治时期。制糖成了当地收益颇丰的生

中世纪时期的埃及糖磨坊

大约 14 世纪西西里的糖磨坊

意，提尔（位于现在的黎巴嫩）就是个著名的糖贸易城市。提尔的威廉曾经编写耶路撒冷王国历史，称糖为一种珍贵的产品，“对人体的健康大有裨益，被商人带到了世界最遥远的角落”。近东的士兵和朝圣者知道了糖，并把它带回自己的家乡。这激发了欧洲人对于糖的渴求——至少皇室和一些贵族对这种甜味剂青睐有加。

10 世纪以来，意大利的城邦威尼斯就一直从东地中海进口和再出口糖。在十字军于 1095 年异军突起之后，这变成了一笔利润颇丰的买卖。于是，威尼斯人扩大了对克里特一带的统治，影响力甚至波及塞浦路斯这样的岛屿。部分得益于蔗糖再出口业务，这个小城邦迅速成了地中海地区最重要的势力之一。虽然热那亚后来成为葡萄牙人从大西洋诸岛运糖的集散中心，威尼斯对于地中海区域蔗糖贸易的控制还是延续了将近五百年。

中世纪时期，一个阻碍欧洲蔗糖产业扩张的严重问题就是劳动力的短缺。地中海蔗糖产区持续不断的战争更加剧了这一问题，而14世纪40年代感染了大半个欧洲的黑死病（鼠疫）更是让情形雪上加霜。在接下来的几十年里，据估计欧洲大约有30%~60%的人口离世，造成了严重的劳动力缺口。与此同时，这个时期还出现了从农村地区向城市流动的人口迁移，这也加剧了甘蔗种植区的劳动力缺乏。西西里岛和其他地中海岛屿的种植园主会给农民付高额的工资，很多欧洲人都前赴后继地来这里工作。然而，干活的人手还是不够，种植园主只好去买奴隶。基督徒和穆斯林都会用奴隶来种植、收获和加工蔗糖。最开始，这些补充劳动力主要是从今天的保加利亚、土耳其和希腊俘虏来的战时囚犯，但后来变成了从东非买来的奴隶，再之后是从西非买来的奴隶。

除了越来越少的劳动力，地中海的蔗糖制造业还有另一个严重的限制因素——气候。甘蔗更喜欢热带气候。寒冷，甚至只是一段时间的凉爽天气，都会阻碍甘蔗的生长。另一个更大的麻烦是这里缺乏廉价、充足的燃料，以便于加热锅炉，使得蔗糖汁转化为精制糖。在地中海产区，对柴火的需求造成了整个甘蔗种植区大面积的森林砍伐。而过度砍伐的连带效应是土壤肥力下降和可利用水资源减少，因为雨水很快流走，侵蚀了裸露的土壤。从15世纪开始，地中海东部——黎巴嫩、叙利亚、埃及和巴勒斯坦——

的制糖产业开始萎缩。到了 15 世纪末，这些地方已经需要进口糖了。在威尼斯人的控制下，塞浦路斯和克里特岛的制糖产业还很兴盛，在地中海西部地区又持续了一个世纪，后来也开始衰退。

而影响东地中海蔗糖贸易的另一个因素，是奥斯曼土耳其帝国的崛起。在 1453 年，土耳其人占领了君士坦丁堡，也就是拜占庭帝国的首都；他们又陆续征服了中东和北非，并进入东欧。土耳其人掌控了横贯东西的陆路贸易，当贸易中断时，欧洲皇室和上层阶级无法轻易从亚洲进口糖、香料和其他珍宝。欧洲人就开始探索能绕过土耳其和阿拉伯人的路线。

大西洋地区的糖

14 世纪以来，葡萄牙人就开始探索东大西洋，陆续发现了马德拉群岛和附近的圣港岛并且将其变成殖民地。他们在岛屿上建立了甘蔗种植园，从 15 世纪中叶起，蔗糖就由这些地方出口到葡萄牙。在葡萄牙本地没卖出去的糖会继续出口到别国，创造了惊人的财富，这也进一步鼓励当地人进行更频繁的探索活动，建立更多的甘蔗种植园。

同样的，西班牙也对大西洋进行探索，并在非洲西北部沿海的加那利群岛建立了殖民地。这些岛屿的气候非常适合甘蔗的生长，当地的原住民被当成拉磨的奴隶。1500 年开始，蔗糖从加那利群岛出口运往西班牙。但和地中海的情况相似，缺乏燃料是个大问题。岛上的森林被过度砍伐之后，制糖业就衰落了，后来因为竞争不过其他地方价格更低的蔗糖生产商而崩溃了。

理想的甘蔗种植地是在热带非洲海岸附近几内亚湾无人居住的圣多美岛和普林西比岛。这些地方是在 1470 年被葡萄牙人发现的。这里有适宜的气候、易得的非洲奴隶、丰沛的灌溉水源和充足的燃料来确保磨坊的运转。蔗糖产业应运而生，尽管将糖运回葡萄牙的航道漫长而曲折，但这笔生意还是给种植园主带来了可观的利润。

［第二章］

从新大陆到1900年

克里斯托弗·哥伦布对大西洋岛屿和那里蓬勃发展的蔗糖产业非常熟悉。作为一家意大利公司在热那亚的代理商，哥伦布曾在1478年赴马德拉采购蔗糖。他第一任妻子的父亲就是圣港岛的执政官。在哥伦布的妻子去世后，他重新娶妻，这位女子的家族在马德拉岛拥有一座蔗糖庄园。当哥伦布第一次航行到加勒比地区后返回西班牙时，他十分确信甘蔗能在他新发现的岛屿上长得很好。1493年他第二次驶向加勒比，途中停驻在加那利群岛，拿了一些甘蔗种子，并把这种作物引入了加勒比海上的伊斯帕尼奥拉岛（现在的海地共和国和多米尼加共和国）。哥伦布和其他西班牙探险家在其他岛上建立了据点，例如波多黎各（1508）、牙买加（1509）和古巴（1511）。甘蔗在这些岛屿上广泛种植，在这之后，同样的事情在西班牙和其他欧洲人建立的中南美洲殖民地上反复上演。

伊斯帕尼奥拉岛是新大陆最重要的蔗糖生产地。从1516年起，蔗糖从这个岛屿出口到西班牙。在此后的30年间，岛上拥有了“动力磨和四匹马驱动的磨盘”。当时的岛屿历史记录者冈萨洛·费尔南德斯·德·奥维耶多·瓦尔德斯写道，西班牙商船运来的“装在货箱里的蔗糖、那些弃之不用的残渣和糖蜜都足以让一个省变得富有。”

虽然加勒比地区有适合甘蔗生长的气候、足够的燃料与水资源，却缺乏人力资源。没什么西班牙人愿意移民到新大陆，在甘蔗种植园里卖力气。像泰诺人和加勒比人这些原住民部落也没什么兴趣在种植园工作。在西班牙人奴

克里斯托弗·哥伦布没有留下一幅真正的图像。这幅版画是在1892年绘制出来，用来庆祝他发现美洲大陆四百周年的。

役他们之后，他们就变得不怎么勤奋了，这是可以理解的。在第一位欧洲人踏上这片领土之后的一个世纪里，不断的战火和欧洲人带来的传染病使岛上大约失去了80%~90%的原住民人口。加勒比地区的蔗糖产业变得萎靡不振。

而巴西就是另一个故事了。葡萄牙人在1500年登陆，不久就建立了小型的沿海贸易港。这里也是种植甘蔗理想的地方：气候很合适，还有充足的锅炉燃料、大量的水资源以及取之不尽的土地。当地的原住民提供了大量潜在的奴隶劳工资源。到了1520年，叫作“engenhos”（这个词在葡萄牙语里是“磨坊”的意思，但被用来称呼一整套甘蔗种植园设施——田地、磨坊和工厂）的小型甘蔗种植园在海岸附近建立起来。到了1548年，伯南布哥已经有6个运转着的甘蔗园了；1583年时，这个数字涨到了66个，另外36个建在附近的巴伊亚，还有其他的建在南部地区。

葡萄牙甘蔗园主因发明和普及了几项非常重要的改进技术而受到赞誉。在17世纪初，这些种植园开始采用一种新的研磨技术，把甘蔗秆放在三个垂直安装的滚筒或者圆柱体之间压碎。甘蔗先是被塞进一侧的两个滚筒之间，然后工人把甘蔗从另一侧的滚筒里翻转过来。相比于传统的挤压式磨坊，这个过程效率更高，所以前者很快就被弃用了。新式的磨还很容易就能通过牲畜、水流，有时甚至是风来驱动。磨坊对劳动力的需求降低，也就意味着蔗糖的产量提高了。

另一项重要的技术革新发生在提纯蔗糖的过程当中。传统的蔗糖磨坊只有一口大锅，里面的甘蔗汁要一直煮到过饱和才行。巴西人发明了一种由多个锅组成的“多釜系统”，液体从一口大锅中被舀到三只容量依次递减的小容器中。这让看着锅的工人对整个过程有了更强的掌控力，也让他们能进行更大规模的生产操作。

巴西的蔗糖产量迅速增加，但当本地劳工资源萎缩时，该行业遭遇了重大挫折。疾病和战火使当地的人口大量减少，巴西的天主教堂开始反对奴役原住民。一条“妙计”横空出世：当殖民地圣多美岛上的葡萄牙蔗糖产业难以与巴西蔗糖产业竞争时，他们开始转型，做起了向巴西出口非洲奴隶的生意。最开始，很多奴隶都是在圣多美岛甘蔗园里劳作的熟练工人。后来，几乎任何在非洲被抓住的人都会被当成奴隶，圣多美岛只是葡萄牙船只的一个停靠点和出发点，这些船只在大西洋上往返，将奴隶转运到巴西和新大陆的其他地方，再把那里的蔗糖带回位于欧洲的家乡。仅 17 世纪这一百年，就有大概 56 万非洲奴隶被运到巴西和欧洲国家在美洲的其他殖民地。

巴西的蔗糖产业被不断强化，大多数产品都出口到欧洲。到了 16 世纪末，糖成了巴西最重要的出口商品，超过了大西洋地区其他产品的总和。在被巴西的产量超越之后，地中海的蔗糖工业就全部消失了，影响力也在大西洋岛屿一带大幅下降。巴西成为蔗糖产业的世界霸主。

糖的提纯

欧洲人将甘蔗种植、蔗糖制造和提纯分成了独立的工作。种植和基础加工都是在大西洋和美洲的殖民地上进行的，而提纯则要在欧洲的城市中完成。制造和提炼的分化有不少好处。首先，这意味着殖民地种植区不需要当地工厂来完成提纯过程，这需要工厂投入大量资金。建在欧洲大城市中心的提纯工坊离最终的市场也更近一些。其次，用船从热带地区运输糖的速度很慢，也很难避免糖在运往母国的路上发生变质。运输不那么精制的糖反而可以尽量降低损耗的风险，而且也可以让欧洲的炼糖厂商生产出顾客所需要的产品。最后，在欧洲城市完成提纯也能为母国带来一些经济效益，而不是全便宜了殖民地。最后一点恰巧反映了当时的经济哲学理念——重商主义。在持这种观念的欧洲人眼里，殖民地不仅是给母国提供原材料的地方，还是重新倾销本土加工产品的市场。

16世纪，欧洲顶级的蔗糖提纯城市是安特卫普。最开始，它控制了来自葡萄牙和西班牙殖民地的蔗糖贸易和提纯。得益于蔗糖产业，安特卫普成为欧洲最为富裕、规模第二大的城市。这样的统治地位一直到1576年才被攻进城门的西班牙精兵打破。安特卫普的蔗糖贸易自此崩溃，城市的影响力也一落千丈。像伦敦、布里斯托尔、波尔多和阿姆斯特丹等欧洲城市进入这个空白市场。它们纷纷开展蔗糖提纯业务，财富也跟着滚滚而至。

加勒比蔗糖业

巴西的蔗糖产业称霸大西洋世界的蔗糖贸易长达一个世纪，然而17世纪中叶，在英国、法国和荷兰殖民者在美洲大陆建成甘蔗种植园后，巴西的市场份额便开始下降。荷兰人在南美洲的北岸建立了甘蔗种植园，也就是后来的苏里南和库拉索岛。1630年，荷兰人占领了伯南布哥的累西腓以及葡萄牙在巴西的其他殖民地，持续了24年的统治。荷兰人允许西班牙裔的犹太教徒在这些地方继续生活，还可以公开进行宗教仪式。荷兰人和犹太人开始熟悉甘蔗种植和蔗糖生产的工作。在葡萄牙人重新夺回荷兰人占领的巴西地区之后，荷兰人和犹太人都被迫离开了这里。有的移居到了英国的殖民地——巴巴多斯。

巴巴多斯在1627年英国建立殖民统治之前是独立的领土。最早的殖民者也不过是一些靠种植烟草来牟取财富的小农。不幸的是，弗吉尼亚和其他一些殖民地以更低的成本生产了更多的烟草。荷兰和犹太裔难民把甘蔗带到了巴巴多斯，还教会了种植园主如何把甘蔗制成蔗糖。奴隶们被从非洲运来种植甘蔗并快速建好了磨坊。制糖产业在这个岛屿迅速发展起来，圣基茨岛、背风群岛和1655年被英国征服后的牙买加也都采取了相同的策略。

法国的殖民过程也差不多，他们在1635年的时候在马提尼克岛和瓜德罗普岛建立了蔗糖殖民地，还在伊斯帕尼奥拉岛的西部建立了种植园。1697年，西班牙和法国签订了《里斯维克条约》，两国瓜分了伊斯帕尼奥拉岛的土地。

18世纪，安提瓜岛甘蔗田里耕种和锄地的景象。

18 世纪，马提尼克岛上的甘蔗种植园。

在接下来的一百年里，法属圣多明各（今天的海地）成了加勒比地区蔗糖产量最高的岛屿。

大型的甘蔗种植园出现在英属西印度群岛。种植者依靠向英国和英属北美殖民地出售糖蜜和朗姆酒覆盖成本。通过出售这些副产品，岛上规模惊人的蔗糖产业赚的几乎是纯利润。一些种植者在制糖业上早已赚得盆满钵满，他们雇用监督者来管理他们的种植园，而自己则乘船回到英国家乡，在那里购置大面积的地产。蔗糖还给英国的商人、提纯工人、船员、银行家、保险家、投资人和蒸馏工人带来了财富。到 1760 年，仅仅布里斯托尔一个城市就有 20 家炼糖厂，每年加工 83.16 万磅（约 37.75 万千克）蔗糖。英国糖业巨头和他们的同盟形成了一股强大的政治力量，从 18 世纪到 19 世纪初一直影响着议会。然而，他们自身经济利益的诉求已经和英属北美殖民地上同僚的意见产生了分歧。一场由糖蜜挑起的经济和政治矛盾悄然而至。

《安提瓜岛十景》（1823）中奴隶砍甘蔗的景象

糖蜜与朗姆酒

为了支撑西印度群岛快速膨胀的奴隶人口，英国殖民者需要进口大量的食物和其他生活必需品，这些东西大多都是从英属北美殖民地运过去的。作为回报，糖蜜、粗糖和朗姆酒也从英属西印度群岛运出。糖蜜是蔗糖提纯过程中的一个副产品，也是比结晶白糖更便宜的甜味剂。它还能用来做酒精饮料，种植园主和商人也会用它来制作高品质的朗姆酒，再出口到英国或者运到非洲换奴隶。

法属加勒比诸岛也会生产朗姆酒，但法国的白兰地制造商对进入法国大都会的朗姆酒进行了抵制。所以，法属西印度群岛的制糖业以处理不了大量过剩的糖蜜而告终。法国政府认为，与其倒到大海里，不如让殖民者将这些糖蜜卖给所有有意向购买的顾客。很显然，他们的目标市场就是英属北美殖民地。

因为法属西印度群岛的糖蜜成本要比英属西印度群岛的低60%~70%，新英格兰殖民者的货船会从马提尼克岛、瓜德罗普岛和法属圣多明各成批地装走糖蜜。新英格兰地区很适合酿朗姆酒，因为那里有掌握蒸馏技术的熟练工人、大量运输庞大糖蜜的船只，以及很多能用作蒸馏燃料和制作酒桶的木材。朗姆酒很快成为北美最受欢迎的蒸馏饮料之选。到了1700年，新英格兰从法国殖民地进口的糖蜜比从本国殖民地购买的还要多。北美的商人往往会拿木材、鱼（主要是给奴隶吃的盐渍鳕鱼）和其他物资来交换糖蜜和原糖。

而这种贸易的结果就是英属西印度群岛的甘蔗种植者失去了大量生意，于是从1716年开始，他们开始敦促英国议会，企图限制新英格兰从法国和其他加勒比地区的欧洲殖民地进口蔗糖和糖蜜。他们拟定的法律使英属西印度群岛在糖蜜和蔗糖贸易中成为一个彻头彻尾的垄断者，允许甘蔗种植者自行定价，这能给他们带来极为可观的收益。议会最终在1733年通过了《糖蜜法案》。该法律宣布向非英国殖民地进口的糖、糖蜜、朗姆酒和其他酒类征收每磅六便士的税。

如果《糖蜜法案》被强制执行，新英格兰地区的渔业和木材生意都会受损，因为这些货物主要就是卖到西班牙、法国和荷属西印度群岛。强制法还会阻碍一些新英格兰人参与其中的奴隶贸易。然而《糖蜜法案》的通过和最终的执行完全是两回事。法案中唯一一条强制执行的条款就是收税的人必须是英国本地的海关官员，而他们大多是糖贸易商的朋友。海关官员通常就只有屈指可数的几位，很容易就能贿赂他们，让他们对在新英格兰海岸线上几千个海港之中运输糖蜜的走私船睁一只眼闭一只眼，允许其悄无声息地卸货。

在18世纪30年代，议会开始允许英属西印度群岛的种植者直接与欧洲国家进行糖贸易。种植者的收入提高了，他们也就不会一味地向议会施压，强制《糖蜜法案》的落实了。即便如此，这个法案还是在英国法典里继续存在了30年，在此期间，糖蜜几乎是被公开走私进北美洲的。不管怎么说，出台《糖蜜法案》又置之不理是一个重要的失误——这个疏忽在后来引起了不小的余波。

美国的蔗糖业

1725年，纽约建成了当地第一家蔗糖提纯厂，加勒比地区的蔗糖制造商开始向这里源源不断地运送原糖。这种半提炼的蔗糖比糖蜜更甜也更贵，但与后者不同的是，它可以被提炼成纯净的糖。更多的炼糖厂在纽约拔地而起，成为城市中最高大的建筑物，提炼蔗糖也成为这座城市最赚钱的产业之一。

七年战争（1756—1763）之后，英国在北美获得了大量的新领地——但要保住这些领地，需要投入很多持续性的费用。为了起到开源的目的，英国议会通过了《食糖法》，降低了30年前设置的对于美洲殖民地进口糖蜜所征的关税。因为这是一个减少现行税金的法案，英国议会中的任何人都没想到这会在美洲殖民地引起什么风波。但这条法案还包括了强有力的执法条款，例如派遣英国海军舰艇在海岸巡逻以及派遣英国海关官员去美洲港口强制征税。

在《食糖法》通过的时候，新英格兰人已经走私糖蜜等违禁物品长达30年了。强制法案让走私变得更加危险。北美殖民者对《食糖法》进行了抗议，而且为了表示支持，很多波士顿和纽约的商人都同意不再从英国进口商品。这条法案迫于压力被废除了，但殖民地的抵制让英国议会出台了更多强制征税的法案来宣示其在北美的主权，而这又进一步激起了殖民地的抗争。最终导致了1775年军事冲突的爆发。

因为英国海军星罗棋布地散在大洋上控制着海洋，北美殖民地的蔗糖贸易和提纯工业在战争爆发之后就瓦解了。美洲蔗糖提纯中心——纽约，被英军占领了八年，蔗糖制造业也被摧毁了。战后不久，蔗糖贸易就被重新启动。纽约的蔗糖提纯工业迅速复兴，随着来自加勒比地区的原糖大量涌入，产业规模不断扩大。1803年，美利坚合众国从法国那里买到了路易斯安那的大片土地。18世纪50年代开始，甘蔗在这片土地（在密西西比河三角洲附近）的南侧耕种，但更南边一点的地方还是保留为谷物生长带；雨季很不规律，适合甘蔗生长的时期也比较短——一年到头只有十个月。甘蔗需要在秋天尽快收获，以免寒冷冻坏这些作物。1795年的时候，法国人让－埃蒂安·博雷将海地的奴隶买到了这里，他们熟知甘蔗的种植和加工。有了这些人的鼎力相助，博雷的种植园大获成功，还另建了新的蔗糖提纯厂。到了1812年，这片土地上已经屹立着75个正在运转的蔗糖磨坊了。1817年引进的带状甘蔗让当地的蔗糖业更上一层楼，这种植物比原先种植的品种成熟得更快。19世纪20年代是甘蔗种植业飞速发展的年代，美国人剥削着大量奴隶劳工，而且联邦政府向进口糖类加征关税也对他们大有裨益。得益于蔗糖的物美价廉，美国每年的人均消费量从1831年的13磅（约5.9千克）增加到了19世纪中叶的30磅（约13.6千克）。

整个19世纪，美国产的大多数蔗糖都是在有着理想地理位置的纽约提纯的。当地的港口设施是整个东海岸最好的，方便了从加勒比地区和路易斯安那州运进原糖。纽约本身就对蔗糖有庞大的市场需求，而且这里完备的公路、运河以及之后的铁路交通网络使精制糖可以便捷地运往北部、南部和西部。曾经在伦敦蔗糖提纯厂当过学徒的德裔威廉姆·哈维迈耶移民到了美国，并于1799年在纽约市运营艾德曼海员公司制糖厂。六年后，他就和兄弟一起开了自家的制糖厂。在那个时候，这只是城里为数不多的蔗糖提纯工厂中的一个——当然，其他的工厂很快就建起来了。1864年，哈维迈耶家族在长岛的威廉姆斯堡建立了世界上最大、技术最发达的蔗糖提纯工厂。

在制作工艺得到改进后，美国的蔗糖产量大大提升，但糖价随之陡然下跌。于是，哈维迈耶家和其他七个制糖工厂引领者在 1887 年一起成立了糖业托拉斯；他们的目的是通过减产来提高糖价，让各家公司获得的利润高一些。随着越来越多公司被并购，这个大型企业被命名为美国制糖公司。在这家公司的规划下，效率低下的工厂被关停，有一些被合并，虽然没什么官方名目，却最终成功将精制糖的价格固定下来。1900 年的时候，这家公司还开设了一个子公司——多米诺糖厂，来推销母公司的精制糖。到了 1907 年，美国制糖公司掌握着全美精制糖产品份额的 97%。

多米诺糖是 20 世纪美国销量最高的糖类品牌。

蔗糖和奴隶

直到19世纪中叶，整个美洲的蔗糖工业都是建立在奴隶制基础上的。奴隶是从非洲获得的，之后被运到美洲来换取蔗糖，蔗糖再出口到英国，然后船上装载满货物再回到非洲交换奴隶。这些奴隶随后被送到加勒比地区的甘蔗种植园去干活。这就是臭名昭著的“三角贸易”。

奴隶在甘蔗种植园的田地、磨坊和工厂里被逼迫做长时间的苦力。他们的寿命因为重体力劳动、营养不良、黄热病这样的恶性传染病和匮乏的医疗而缩短。特别是在巴西和牙买加，一些奴隶向内陆逃亡，建立起自己的社群。病死或者消失不见的工人又需要被人替补上，因此，在18世纪前75年间，光是西印度群岛就吸纳了120万非洲奴隶。据估计，在1700至1810年间，大概有252000名奴隶被送到了巴巴多斯，另有622400名奴隶抵达牙买加。奴隶人口在数量上很快就超过了欧洲人，特别是在那些加勒比海的岛屿上。在1789年的法属圣多明各，仅由32000人的少数白人控制了大概50万名奴隶。

奴隶起义时常在巴西和加勒比地区发生，但通常都会被暴力镇压，起义者会被用极为残忍的方式处死。唯一一次成功的奴隶起义发生在1791年的法属圣多明各。那里的奴隶受到法国大革命和宣称人人生而平等的《人权宣言》的启发，反抗了他们的奴隶主。法国甚至派遣了军队到岛上镇压反叛，但士兵败在黄热病和对手的游击战策略之下。1803年，

19 世纪初奴隶起义后残留的海地蔗糖工厂遗迹，大约摄于 1830 年。

起义终于获胜，1804 年 1 月 1 日，海地（原法属圣多明各）成为美洲第二个独立的民主国家。在整个起义的过程中，没有被立刻杀死的白人庄园主和监工都逃到了其他的殖民地，一些去了路易斯安那，一些去了古巴。然而，曾经在加勒比地区最发达的海地的制糖工业却再也没有恢复。

在 18 世纪末，欧洲和北美兴起反对奴隶制的运动。贵格会教徒和其他一些人开始禁食由奴隶生产的糖类制品，但这些策略都还仅限于分散的个人层面。1791 年，英国议会没能成功通过废止奴隶贸易的法案，于是，英国那些反对奴隶制的人就加入了抵制奴隶种植的甘蔗制糖的行列。

由于糖通常会被端到家庭茶几上食用，因此女性在废奴运动中起到了尤为关键的作用。这个运动发展了很多支持者，人数高达 40 万。反对奴隶制的人不仅仅是那些废奴主义者，还有自由主义经济学家，如《国富论》（1776）的作者亚当·斯密就指出，使用奴隶的成本远高于其创造的经济效益。还有的人是担心西印度群岛种植园主不正当的政治权力，毕竟整个 18 世纪他们都在操纵议会，制定符合他们利益、损害英国本土经济体的法令。

英国愈演愈烈的废奴运动鼓励人们购买从印度进口的、非奴隶制造的蔗糖。起初，没什么印度产的糖被运到英国，但随着废奴运动发展壮大，订单越来越多，到了 19 世纪初，在英国买到印度糖已经不是什么难事了。贵格会教徒还创办了“自由生产协会”，贩卖亚洲生产的糖。

艾萨克·克鲁克香克，“奴隶贸易的逐步废止——又名，慢慢戒糖”，1792 年。

EAST INDIA

SUGAR BASINS.

B. HENDERSON,

China-Warehouse,

RYE-LANE, PECKHAM,

Respectfully informs the Friends of Africa, that she has on Sale an Assortment of *Sugar Basins*, handsomely labelled in Gold Letters: "*East India Sugar not made by Slaves.*"

"A Family that uses 5lb. of Sugar per Week, will, by using East India, instead of West India, for 21 Months, prevent the Slavery, or Murder of one Fellow-Creature! Eight such Families in 19½ Years, will prevent the Slavery, or Murder of 100!!"

PRINTED AT THE CAMBERWELL PRESS, BY J. B. G. VOGEL.

废奴主义者广泛宣传非奴隶制造的印度糖。

在美国，废奴主义者开始尝试种甜菜，并且抵制从加勒比进口蔗糖。美国贵格会教徒还支持将枫糖产业作为另一种选择，从18世纪80年代开始，一小部分工厂开始生产枫糖。1789年，费城人同意以统一价格购买一部分枫糖，来帮这些工厂站稳脚跟。贵格会特别呼吁使用枫糖来“减少成千上万的非洲人为了满足我们的口腹之欲，而在种植甘蔗时承受的痛苦”。贵格年鉴里还建议读者在家做枫糖，因为它比“混着奴隶的呻吟和泪水”的蔗糖更甜。19世纪30年代，《主教录音机》和《有色美国人》等报纸的文章会敦促家长禁止孩子去糖果店买甜食，因为所有购买蔗糖的行为都是在支持奴隶制所犯下的“罪孽”。

这些人在抵制买卖和宣传废奴上下的功夫总算是取得了一些胜利。1807年3月3日，美国总统托马斯·杰斐逊签署了“禁止将奴隶贩卖至美国管辖的任何港口或地方”的法令。三周之后，英国上议院也通过了《废除奴隶贸易法案》。奴隶制在英属西印度群岛一直延续到1834年才结束，在法国殖民地存续到1848年，而美国更是要到1866年才正式废奴。古巴一直到1886年还允许使用奴隶，巴西的奴隶制则是在1888年才结束的。

19世纪末，真空盘、离心机和蒸汽动力在蔗糖提纯中的应用等技术进步都让整个产业降低了劳动强度，但仍然需要大量劳动者。在获得解放后，自由的奴隶不想继续在甘蔗种植园工作。对于劳工的需求让这些工厂不得不雇用契约劳工来种植园工作，这些劳工大多数都是印度人和中国人。因此，成千上万的合同工涌入这些种植甘蔗的地方，在合同结束后，很多人就留在了当地。

古巴蔗糖业

法属圣多明各的蔗糖业终结之后，最直接的受益者就是西班牙殖民地古巴。自1523年起，甘蔗就在这片土地上生长，然而，因为西班牙禁止古巴人直接同外国商船交易和限制进口奴隶的法律条款，岛上的蔗糖业一直不怎么成熟。

古巴的蔗糖业直到1762年才开始发展，那时候英国因为七年战争对哈瓦那进行了为期10个月的统治。在这期间，英国人带了成千上万名奴隶来到古巴。战争结束后，英军就撤退了，古巴蔗糖生产者开始要求更自由开放的政策。西班牙放宽了法律，对奴隶买卖和同外国船只贸易的活动开了绿灯。在18世纪80年代，超过18000名奴隶被带到了古巴，18世纪90年代以及19世纪的前十年里，超过了125000人。古巴蔗糖业因此变得兴盛起来，大多数蔗糖都会被出口，以此换来了很多欧洲和美国的加工品。然而，到了1790年，古巴一年的制糖量也不过15000吨。但惊人的变化已经悄然而至了。

在法属圣多明各的奴隶起义期间，那个岛上的法国人带着他们的奴隶搬到了古巴，在这里建起新的甘蔗种植园和工厂。得到改善的运输系统也让古巴的制糖业更进一步；新的公路和之后的铁路让提纯厂生产出来的蔗糖能很方便地运到出口的海港。同时期发生的则是加勒比岛屿上蔗糖的减产，主要是受到英属、法属殖民地上轰轰烈烈的废奴运动的影响。古巴保留了奴隶制，很快跻身全世界成本最低廉的蔗糖制造商之列。蔗糖成了古巴最主要的出口农副产品，美国成为当地人的重要贸易伙伴。小的蔗糖磨坊被关停了，效率更高的集中化工厂服务于几位大种植园主。在美国南北战争期间（1861—1865），路易斯安那州的甘蔗种植园遭受重创，世界市场上的糖价暴涨，而古巴蔗糖借此机会占领了更多的市场。从 19 世纪 40 至 70 年代，古巴所占的世界蔗糖市场份额从 25% 一跃上升到 40%。

然而，岛上 1868 至 1878 年之间发生了奴隶起义，使古巴蔗糖业陷入停滞。在战争期间，很多蔗糖生产商离开了古巴，在多米尼加共和国做起了小买卖。在古巴奴隶于 1886 年获得解放之后，很多人逃离了种植园，不愿意再从事蔗糖制造相关的劳动。古巴便向合同劳工抛出了橄榄枝。在接下来的十年间，这个国家大约吸纳了来自西班牙、美国、中国、海地和其他加勒比岛屿的共 120 万移民。

还有一大麻烦，就是来自海外的甜菜的市场竞争日益激烈，甜菜往往是在欧洲和美国种植并转化提炼成糖的。好在大型的美国公司开始大规模向古巴蔗糖提纯工厂注资。1890 年，这些商业巨头游说美国国会颁布了《麦金莱关税法》，取消了对古巴精制糖征收的进口关税。1896 年，光是美国糖业托拉斯一个组织就掌握了 19 家古巴蔗糖提炼厂。

古巴对美国的蔗糖出口量暴增，而古巴从美国进口的

古巴工厂加工甘蔗的景象，出自《古巴商业：写给商人的书》（插图版）（1899）。

19世纪的古巴蔗糖工厂

加工品数量也节节攀升。1894年古巴的蔗糖产量达到了110万吨。但后来美国的甜菜种植者和蔗糖提纯业者成功游说国会，对从古巴进口的蔗糖征收40%的关税。西班牙人对从美国出口到古巴的货物也报复性地征收了一笔关税。古巴原糖的价格急剧下跌，而从美国进口的商品价格却飙升。甘蔗种植园的合同工被辞退了，许多人加入了古巴游击组织，开始了摆脱西班牙统治的独立运动。游击队摧毁了很多蔗糖提炼厂和种植园，西班牙殖民政权则实施了很多强硬的举措加以回击，包括建立集中营关押起义军来镇压。然而，这些暴行被很多美国媒体揭露；那些极具煽动性的、被称为“黄色新闻”的文章左右了美国民众的舆论，让他们站在了古巴游击队员的一方。

美西战争及其余波

1898 年 2 月，美军舰艇“缅因”号在哈瓦那港口爆炸沉没。尽管原因并未确定，但美国人却将这笔账算在了西班牙人的头上。在爆炸发生两个月后，美国向西班牙宣战。在为期五个月的军事冲突期间，美军占领了古巴、波多黎各、关岛和菲律宾。美国还吞并了夏威夷，后来这块土地为美国糖业集团所用。

战后，波多黎各、夏威夷和菲律宾的蔗糖产量有了一定规模的增加，古巴在获得了水力磨坊、密闭炉、蒸汽机和更先进的真空盘技术之后，蔗糖产量直线上升。美国人在古巴蔗糖业的投资也越来越多。到 1919 年，美国人掌握了当地 40% 左右的蔗糖工业。古巴的蔗糖产量在 1925 年达到了 350 万吨。

20世纪初，收割甘蔗的景象。

回溯历史

在哥伦布1492年首次驶向加勒比海之后的四百年间，蔗糖业经历了天翻地覆的变化。产业中心从地中海和大西洋岛屿转移到了美洲。种植园里的主要劳动力也从奴隶变成了契约劳工。甘蔗的收获、研磨和加工等工序从大量依靠手工劳动升级为一个基于机械和科学发明的最新技术的工业化系统。蔗糖的生产也从小的种植园和磨坊转向以跨国公司为基础的生产。这些变化带来了世界范围内的蔗糖价格的直线下跌以及糖消费量的急剧上升。

[第三章]

糖的全球化

甜菜（甜菜属）原产于地中海，早在新石器时代它的根和叶子就在欧洲和中东被广泛食用。古希腊人和罗马人会把甜菜种在自家的花园里。行医的人把它作为治疗各种疾病的药物。在中世纪，甜菜作为一种花园植物存活，到了15世纪，欧洲几乎随处可见它的身影。16世纪的草药志里记录着几种甜菜品种，其中就包括有甜味的浅色甜菜。在法国农业学家奥利维尔·德·塞雷斯所著的《农业戏剧》（1600）中，这位农学家首次观察并记录下甜菜根“可以作为食物，其烹饪时产生的汁液就像糖浆一样”。

甜菜是种生命力顽强的作物。不像只能生长在温和气候下的甘蔗，这种耐受度极高的植物能抵御干旱和洪水。它们的生长期也比较短，收获之后地里还能种点别的作物。根部（甜菜根）可以风干保存之后再食用，也是牛和马理想的饲料。在17世纪，它们已经是欧洲大陆上很普遍的农作物了。

甜菜根另一个很重要的特性是一位普鲁士化学家安德烈亚斯·S. 马格拉夫发现的。1747年，马格拉夫向位于柏林的科学院提交了论文，宣称自己从甜菜根里提取到了少量的蔗糖。事实上，马格拉夫用于实验的甜菜品种产糖量很低，而且他的提取方法既不实际又不划算。但这些都是瑕不掩瑜的小事，只要这个过程可以被改进，非热带国家不需要进口就能实现糖的自给自足了。在后续的几十年里，普鲁士政府陆陆续续地对甜菜根糖生产研究提供了资金支持。

马格拉夫在1761年重复他的实验，生产了足够的糖来制作一些面包，但从商业的角度看，整个过程还是不太实际。在他1782年逝世后，他的学生弗兰兹·卡尔·阿查德继续进行与甜菜有关的实验，并且发现一些品种比其他品种产糖量更高。他在1799年进一步改善了马格拉夫的提炼法，并献给了普鲁士国王弗雷德里克·威廉姆三世几磅甜菜糖结晶。两年之后，国王给了阿查德一些资金支持，让他在西里西亚建立工厂，试验他的方法。阿查德学到了很多关于甜菜的知识，他被认为是第一个在商业化基础上实现从甜菜中提取蔗糖的人。他发现白甜菜的蔗糖含量最高，随后他用这种甜菜进行育种。阿查德声称，在国内生产甜菜糖要比进口甘蔗糖划算得多，但他的工厂却并不怎么成功。

甜菜制糖产业在拿破仑战争时期复兴，当时法国实施了“大陆封锁政策”，阻止来自英国或其殖民地出产的货物进入被法国控制或者是与法国结盟的欧洲国家。这些货物中就包括遍布欧洲、从英属西印度群岛进口的蔗糖。1791年，法属殖民地圣多明各爆发的奴隶起义又使法国在加勒比地区殖民地用于出口的蔗糖量锐减。战争期间，英国封锁了法国控制的陆上港口，使得糖很难从任何地方进口。于是，法国为甜菜糖生产提供了奖金，随后100多家甜菜糖工厂建立起来，主要坐落于法国北部，也有些散落在欧洲大陆的其他地方。虽然蔗糖被顺利提取出来，但当1815年恢复和平的时候，廉价的蔗糖再次从加勒比海地区流入，这里的制糖业崩溃了。

然而，甜菜从来没有被人们遗忘。法国的种子生产者

弗兰兹·卡尔·阿查德（1753—1821）

和植物选择性育种的先驱维尔莫林开始了育种实验，以提高甜菜的含糖量。1837 年，有更高蔗糖含量的品种糖用甜菜 (*B. vulgaris var. altissima*) 问世了。几乎是在同一个时间，新的提取技术被发明出来，降低了生产甜菜糖的成本。德国、法国、比利时、奥匈帝国、俄罗斯和斯堪的纳维亚的甜菜制糖工业复苏。最终，育种人能培植出根部含糖量高达 20% 的甜菜品种。

即使研制出了含糖量更高的甜菜，发明了更先进的提取技术，进口蔗糖仍然比提取甜菜糖更便宜。最开始支持甜菜糖的是贵格会教徒和废奴主义者，他们反对由奴隶生产出来的甘蔗糖。然而，随着加勒比地区的奴隶在 19 世纪中叶得到了解放，支持甜菜糖的声浪也就在英美地区偃旗息鼓了。在加勒比地区的奴隶制终结之后，糖价理应上涨，

阿查德在西里西亚的甜菜糖工厂

但甜菜和甘蔗的生产在世界范围内不断扩大，使得19世纪的糖价一直呈下降趋势。各国政府一直致力于支持甜菜制糖产业，通过对进口蔗糖征收关税和进口配额，颁布了很多有利于本土甜菜种植者的政策。有了政府的干预，欧洲的甜菜种植不断发展，在19世纪末新建了数百个工厂。这样的政府支持在20世纪还在延续，使欧洲在20世纪末成为糖的净出口方。

美洲也种植甜菜，那里的人们早在19世纪30年代就开始尝试把甜菜转化成糖。然而，美国直到19世纪70年代，而加拿大直到80年代，才实践成功。有了政府的保护和支持，甜菜糖产业在20世纪初迅速发展。机械收割等技术的进步提高了产量和效率。

拿破仑因为支持甜菜工业而受到鄙夷。

20 世纪 30 年代左右，成熟的甜菜田。

1915 年，小朋友在科罗拉多州糖城的甜菜地里劳作。

从托马斯·柯南特的《上加拿大速写》（1898）中不难看出，枫糖浆仍是加拿大和美国的重要甜味剂。

非洲、亚洲和大洋洲的糖

当甜菜种植在世界温带地区发展的时候，甘蔗种植和蔗糖制造在非洲、亚洲和太平洋的热带地区扩大。毛里求斯——英国在拿破仑战争期间获得的一个印度洋上的殖民地——有着适合种植甘蔗的热带气候。甘蔗在1829年被引入了这片土地。然而，随着产业规模的扩大，它面临着严重的劳动力短缺。英国种植园主的方案就是签订契约劳工，成千上万的印度人随即涌入这里。到19世纪中叶，这座岛屿生产的糖占全世界糖总量的9.4%，随着西印度群岛蔗糖产量的下降，毛里求斯成为英国最主要的供给地之一。在合同期满后，很多印度人没有返回家乡，而是留在了毛里求斯。在1975年这个国家独立的时候，大多数人口都有印度血统。毛里求斯继续种植蔗糖，欧盟是其主要买家。

甘蔗还曾生长在纳塔尔，它如今是南非的一部分。随着甘蔗的种植面积不断扩大，纳塔尔附近的祖鲁兰（今称夸祖鲁－纳塔尔省）在1887年被吞并后也成了甘蔗的种植区。同样，种植园主的主要问题是寻找合适的劳动力。非洲人对种植园主提供的工作条件和工资并不满意，园主便想到从印度购买大量劳工，但印度人也发现这个工作令人不愉快，纷纷转去从事别的职业。很多人没有回印度，而是留在了南非做自己的小买卖。印度劳工后来被南非其他地方移民过来的人取代，其中包括很多纳塔尔、夸祖鲁和莫桑比克的童工。

在史前时代，甘蔗就由波利尼西亚人和美拉尼西亚人带到了太平洋岛屿，他们在出发远航的时候会带上甘蔗。当他们到达了新的岛屿，就在那里种下甘蔗。英国人最早在澳大利亚种植甘蔗是在1788年。最开始，他们是打算让甘蔗在悉尼扎根，但那里寒冷的气候实在不适合种植。到了19世纪60年代，制糖业在澳大利亚重启。在昆士兰州，最开始用罪犯来充当劳动力；新南威尔士州则是从波利尼西亚雇用契约劳工。昆士兰州开始只有小农种植甘蔗，在19世纪80年代转变为种植园的模式，雇用从美拉尼西亚来的合同工。昆士兰州还把很多林地转变为农业用地，以糖为主要作物。

到1900年，制糖业是美国最为重要的产业之一——其在路易斯安那州、得克萨斯州和新获得的波多黎各、菲律宾和夏威夷，都有广袤的甘蔗田；甜菜则种在西部各州，如犹他州和加利福尼亚州。蔗糖提纯工业在很多东部城市中扮演的角色也更加重要。而其中数一数二的就是哈维迈耶在纽约开办的工厂。之前我们就提过，1887 年亨利 · O. 哈维迈耶和其他蔗糖提纯厂建立了合作，创办了食糖提纯公司，也就是人们平常说的糖业托拉斯。

美国的莱德公司在 1835 年租用了考艾岛的土地来种植甘蔗和建蔗糖磨坊，其他公司也纷纷效仿。然而，几乎没有夏威夷本地人愿意在甘蔗种植园里工作，所以庄园主就在海外寻找更为便宜的劳动力，随即运来了大批契约劳工，最初主要是中国男人（女人是被排除在外的）。到 1860 年，夏威夷一共有 29 个甘蔗种植园，这个地区的大部分蔗糖都会出口到美国。当马克 · 吐温在 1866 年到访这里时，他对当地的甘蔗种植情况印象深刻，称夏威夷“就其惊人的产量而言，它是世界产糖之王”。1875 年，夏威夷和美国签订了互惠互利条约，允许美国免税进口夏威夷蔗糖。

第二年，在加利福尼亚开办甜菜糖厂的德裔移民克劳斯 · 斯普雷科尔斯来到了夏威夷，很快就购买了夏威夷大部分的糖。到了 19 世纪 80 年代，他最终拥有或掌控了加利福尼亚和夏威夷岛上生产的大部分蔗糖。

到 1882 年，夏威夷大概 49% 的蔗糖业工作者都是中国劳工，政治领袖开始担心当地大量的外籍居住者。接下来的一年，夏威夷不再接收来自中国的移民，很多华人也纷

20世纪初，加利福尼亚和夏威夷蔗糖公司的工厂。

纷离开。而在1887年，美国糖业利益集团逼迫夏威夷国王接受宪法，将王国内的大部分权力赋予他们。有了美国海军的支持，欧美商业利益集团于1893年推翻了夏威夷王国的君主政权。他们进而向美国国会施压，想要吞并这些岛屿，这个目标最终在1898年美西战争时期实现。

甘蔗田和蔗糖提纯厂里都需要很多人手。冲绳人、韩国人、波多黎各人、葡萄牙人和菲律宾人都来到夏威夷的蔗糖园里工作，但其中规模最大的移民群体当属日本人，他们从1865年就作为契约劳工来到这里。合约期满后，很多日本劳工还留在夏威夷，时至今日他们的后代依旧是当地最大的单一民族群体。

加利福尼亚和夏威夷蔗糖公司（C&H糖业）自1906年

成立以来就一直主导着夏威夷的蔗糖产业，直到 20 世纪 30 年代甘蔗种植园都被转为他用。今天，岛上仅有一家甘蔗种植园。C&H 糖业现在是美国制糖公司（多米诺糖业）的一部分，该公司归佛罗里达水晶制糖公司和佛罗里达甘蔗种植合作社所有。

佛罗里达算不上一个适合种植甘蔗的地方。那里的亚热带气候时有冰冻期，会给甘蔗作物造成毁灭性的打击。该州南部地区稍微好一些，但大多数地都被大沼泽地国家公园占据。19 世纪末，佛罗里达东部建立了甘蔗种植园和制糖厂。1821 年，佛罗里达被西班牙割让给美国，当地制糖业迅速发展；然而，这些种植园却因为无力匹敌路易斯安那州的本地蔗糖和加勒比地区的廉价的进口糖而陷入停滞。

加利福尼亚和夏威夷蔗糖公司的工厂内，工人在给蔗糖装袋。

直到美国糖业公司于20世纪30年代在克莱维斯顿落成，蔗糖才在佛罗里达州获得了第二次现身的机会。1942年以前，美国糖业公司都是一家效益不温不火的公司；所幸二战带来的大量蔗糖需求使得甘蔗的种植面积扩大，然而，这还只是个开始。1948年，美国陆军工程兵团开始了在大沼泽地国家公园的排水作业，并且修建了灌溉系统来保护佛罗里达州南部人口稠密地区免遭暴风雨的摧残。在此后的50年间，超过一半的沼泽区被排干了水分，开垦的土地成为沼泽地农业区。事实上，这里几乎可以种任何农作物，但当地的气候和政府的补贴都鼓励农民种植甘蔗。

当菲德尔·卡斯特罗在1959年掌管古巴之后，美国大幅度减少了对古巴糖的购入。作为反击，古巴政府将古巴的蔗糖企业国有化，这些工厂原先多是美国人的资产。于是，美国从1961年开始就暂停了从古巴进口蔗糖的全部订单。像范朱尔家族这样富庶的古巴甘蔗园主和制糖业者纷纷逃离古巴，在佛罗里达州的南部购买地产，继续种植甘蔗。范朱尔家族还买下了一些在多米尼加共和国的甘蔗园。

1962年，美国糖业公司在靠近佛罗里达州西棕榈海滩的地方开办了布莱恩特糖业公司。这算得上是世界上最现代化的工厂了。到了20世纪80年代初，美国糖业公司已经是州内最大的蔗糖生产商了，而且佛罗里达还是全美第一制糖州。这样的辉煌部分归功于美国农业部，从1979年开始，他们就为本土农民提供补助和“无追索权贷款”（也就是说不用还本金），还对从多米尼加共和国、巴西和菲律宾进口的蔗糖设定限额。

补助不成比例地落入最大的甘蔗种植者和甜菜种植者的口袋里。1991 年，42% 的补贴金全归 1% 的种植者享有。正是这样国外竞争小、国内有补贴的模式，让美国消费者从 1985 年起为糖支付的价格比世界市场价格高出每磅 8 至 14 美分——这还不算那些本就出自他们税金的补助款。因为糖还是很多再加工产品的原料，这些不正当的补助造成了食物成本的大幅上涨。但不得不说，补助、定额和关税确实促进了美国制糖产业的发展——到 2000 年，美国人吃的原糖的 60% 都产自本土。

得益于北美自由贸易协定，墨西哥从 2008 年起不用付关税也可以很方便地将蔗糖出口给美国和加拿大。这使得墨西哥的甘蔗种植和蔗糖提纯在一夜之间兴起。现在，墨西哥已经是全球第七大蔗糖生产国了，2013 至 2014 年，该国提供的食糖占了美国糖总消耗量的 15%。

战争和革命

在 1898 年美西战争的时候，美军占领了古巴。1903 年，古巴获得独立，两国签订了互惠协定，除了其他条款外，规定将进口到美国的古巴食糖的关税降低 20%。即使仍然需要征收关税，古巴糖还是比美国本地糖卖得要好。在接下来的十年里，古巴成为美国糖的主要外国供应国。于是，像里黑昂达公司和古巴贸易公司这样的美国企业并购了古巴的一些蔗糖厂。1936 年，这些产业的继承人老阿方索·范朱尔入赘了古巴一个知名的蔗糖家族，通过整合资源成为当地最大的蔗糖生产商，也是世界上最大的食糖企业之一。

随着美国甜菜产业的扩张，糖价逐步下跌。1929 年，大萧条来临，国会通过了《斯姆特－霍利关税法案》，通过向进口糖加征关税来保护本土糖产业。古巴蔗糖工业在价格上遭受了沉重打击。1934 年，美国国会通过了《蔗糖法案》(《琼斯－科斯蒂根法案》)，掌控了蔗糖的进口，支持国内蔗糖的生产和提纯。二战之后，古巴蔗糖进口量再次回升，占据了美国食糖总消耗量的 25%~51%。

1961 年，古巴实现了蔗糖企业的国有化，没收的甘蔗种植园成为国有企业。工人被许诺终身的职位，政府向他

古巴种植园收割甘蔗的景象，1904 年。

《冰球》杂志的封面上，“古巴女士”避开递给她一个标有“降低古巴蔗糖关税”盘子的山姆大叔，1902 年。

们施压以实现生产目标；但因为没有什么经济激励政策，产量反而下降了。1968 年，这种作物的收成太惨淡了，以至于政府将收割甘蔗军事化才能完成生产任务。

美国在 1960 年终止从古巴进口蔗糖的贸易之后，苏联和东欧国家接手了这个摊子——在接下来的30年内，他们买走了古巴出产蔗糖的87%。在1991年苏联解体之后，很多古巴制糖厂也宣告破产——在国内156家制糖厂中，71家关门，60%的甘蔗田变成了蔬菜园和牧牛场。然而，在当地人学会用甘蔗制造乙醇（一种可以给车供能的酒精燃料）的工艺后，古巴的甘蔗业又在21世纪焕发了新的生机。

英国糖业

自17世纪中叶以来，英国人就很乐意从地中海和大西洋诸岛进口原糖，再把它们运往伦敦和其他城市提纯。也是在这个时候，英国不再从欧洲其他国家购买蔗糖，它得到了包括巴巴多斯和牙买加在内的西印度群岛的殖民地，一跃成为最主要的蔗糖制造商。19世纪，英国商人在其他不同的地方开展了甘蔗种植和蔗糖提炼的业务，比如印度洋上的毛里求斯、南非的纳塔尔和澳大利亚东北部的昆士兰。

19世纪中期，英国的蔗糖提纯产业被两位商人牢牢掌控——亨利·泰特和亚伯拉罕·莱尔。泰特是利物浦一位成功的杂货商人，后来成为约翰·怀特蔗糖提纯公司在利物浦的合作人。泰特在1869年接管了公司，将其重新命名为亨利·泰特父子公司。他还在利物浦和伦敦银城建立了其他的提纯厂，专做方糖块。

泰特的主要竞争对手就是苏格兰商人亚伯拉罕·莱尔，他在1865年和商业伙伴一起掌控了苏格兰格林诺克的格莱

英国的锅炉房，大约 18 世纪。

图片展示了收割下来的甘蔗是如何在伦敦被提炼成精制糖的，19 世纪 30 年代左右。

贝制糖厂，6 年后在东伦敦建起新的厂房。他的工厂主要生产的是金黄糖浆，一种颜色较浅却香气馥郁的液体甜味剂，可以用来做蜜饯、烹饪食物或者放在餐桌上当餐用糖浆。金黄糖浆在 1904 年被注册成为商标，这大概是英国历史上第一个注册的品牌。1921 年，这两家公司合并为泰莱公司，一直到 2010 年都保有全英最大制糖厂的名号，在这一年他们将其制糖业务出售给了美国制糖公司。

19 世纪末期，英国开始种植甜菜，但这个产业一直到进口蔗糖变得举步维艰的一战时期才正式发展起来。20 世纪 20 年代，这个产业开始了真正意义上的腾飞，但在随后的大萧条时期受到重挫。1936 年，英国将甜菜产业国有化，将几家公司合并为英国制糖公司。1991 年，英国制糖公司成为英国食品联合集团（ABF）的子公司。今天，英国更多的糖是从甜菜里提取的，而不是从进口的甘蔗里。

Buy your Sugar in sealed packets and obtain guaranteed quality with full net weight.

刚刚合并的泰莱公司向消费者提供一些糖制品的使用方法，20 世纪 20 年代左右。

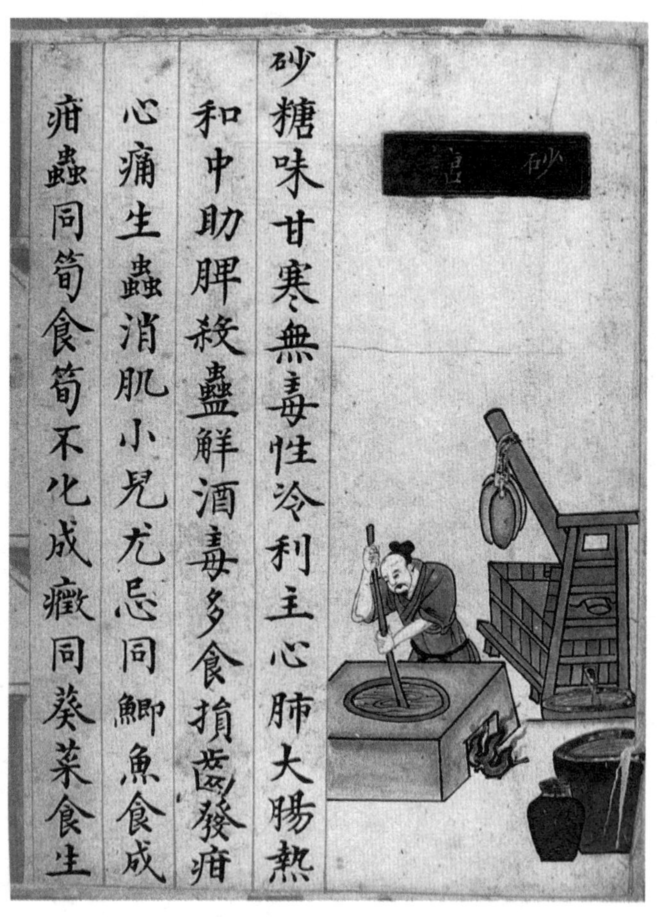

砂糖
砂糖味甘寒無毒性冷利主心肺大腸熱
和中助脾殺蠱鮮酒毒多食損齒發疳
心痛生蟲消肌小兒尤忌同鯽魚食成
疳蟲同筍食筍不化成癥同葵菜食生

《食物本草》中对于砂糖的介绍，这是一种可以追溯到中国明朝（1368—1644）的草药。在男子的身后，是一个可以将甘蔗挤压出汁的压榨机。

[第四章]

糖的用途

依据糖的历史来看，曾经唯一品尝它的方法就是嘬一嘬或者咀嚼甘蔗秆，享用它的甜汁。在甘蔗种植区，人们用这些甜味汁液及其副产品来制作甜食和酒精饮料的历史至少有 2500 年。在古代的印度次大陆，蔗糖会被加到用枣酿成的酒中；果汁也会用糖来调味；糖水有时候会先用其他香草调味，再加进其他饮料里。在“广博仙人”毗耶娑所著的印度史诗《摩诃婆罗多》（公元前 400 年）里就提到了用糖和克里萨拉做成的甜点，后者是一种由牛奶、黑芝麻粉、大米、糖和香料这五种食材制成的流食。虽然成分和黏稠度稍有变化，但是这种组合流传了下来，成了一种叫“punch”的食物（取自“panch”的读音，在梵文里是“五”的意思）。最开始，这些食物和饮品只有在富人家和举办特殊节日庆典的时候才能被享用。然而到了 13 世纪，糖在印度已经足够多了，即使在不太富裕的甘蔗种植区也随处可见。

13 世纪，蔗糖在中国的南部和东部也非常常见。中国文人吴自牧在《梦粱录》中就提到了开在杭州的七家甜食铺子；这些店里会卖彩色的、像花一样的糖果，还有甜米粥、拉丝糖、调味馅料、麝香糖和浸泡在糖蜜里的蜜饯。两本 13 世纪中叶的中国食谱也流传了下来，里面都记载了用甘蔗制品制作糕点、糖果和糖浆的方法。《中国：糖与社会》（1998）的作者穆素洁计算过，这两本书中用到糖的菜谱分别占了全部内容的 17% 和 25%。两者都介绍了用糖保存蔬菜和水果的方法。这的确是个很奏效的法子——蔬果虽然在丰收季来得容易，但一摘下来很快就会腐烂。用糖保存使得人们在得不到新鲜食材的非收获季也能享用它们。糖还能用来掩盖没有熟或者熟过头的水果不好的味道。在中国南方，糖渍蔬菜和水果非常受欢迎，在街边小贩的推车上、茶馆里和酒馆中都会贩售。

制作冰糖需要把蔗糖浆煮沸至过饱和的状态，倾倒进模具里，然后在太阳下干燥。还有一种由糖和磨碎的松子（或者核桃）制成的糊状物，可以压进模具里，制作出可食用的花朵、动物、鸟类和水果形状的糖雕。糖和磨碎的坚果的混合物后来演化为一种杏仁糊，也就是杏仁蛋白糖，是中东和地中海地区标志性的糖果。

甘蔗在 7 世纪来到了中东，迅速成为波斯、伊拉克和埃及餐桌上的美食。伊本·萨亚尔·瓦拉克在 10 世纪写的巴格达烹饪书中就提到了多达 80 种的蔗糖用法，包括制作红酒、糖蜜杏仁和核桃、饼干、脆饼、布丁、牛轧糖和硬糖的食谱。很多食谱都以不同的形式流传了下来。像瓦拉克的牛轧糖（在书里叫“natif”）食谱，很可能是现代土耳其软糖的灵感来源。瓦拉克的书里还提供了专门给小孩、老人和旅行者的食谱。作者还特意注明了糖有一定的药用功效，可以缓解喉咙、胸部和肚子的疼痛，还有一些其他的妙用。

在那个时代，撒满糖和果脯的迷你派很受欢迎。另一种常见的美食是一种折叠了几次的薄脆饼，浸满融化的黄油，再用蜂蜜或糖来调味。库纳法也算一个，这种薄薄的、切得细碎的小面团要用纯净黄油保持湿润，刷上蜂蜜和糖蜜，再拿去烤制。在阿拉伯国家，糖还出现在了一些咸味菜里，比如用桃子、杏和大枣（热带树木金丝小枣的果实，有时候也叫中国枣）一起炖煮的羊肉和羊排。糖也被用来给饮料增加甜味。阿拉伯人会做一种叫果子露的饮品，是将加入了玫瑰花瓣、橙花、柳花或者紫罗兰的糖浆搅入冷水后制成的。果子露也能用葡萄干或者其他的果酱制作。如果在这种糖水中加入冰，它就变成了一种新的冷冻甜点，欧洲人叫它冰冻果子露。

而精制糖在埃及的富人家庭达到了炫耀性消费的巅峰。《伊斯兰文艺复兴》（1922）的作者亚当·梅兹就指出，公元970年，卡福尔这个维齐[①]家庭每天会吃掉1000磅（约450千克）的糖；在之后的一个世纪中，有一位埃及维齐的宴会上摆出了20吨重的糖雕，有城堡和各种动物的形象，如大象、狮子和鹿。而他的另一场宴会用了5万件糖雕做装饰，每个都有大概4磅重（约1.8千克）。

糖在欧洲的使用

在9世纪以前，少量的糖被运到了欧洲，主要是当作药物来使用。在那个时代，“体液学说”牢牢占据着欧洲医药理论的主流长达一千年；当时人们认为，人的健康和脾性

① 维齐（Vizier）是对古代伊斯兰国家高官的称呼。——译注

都与体内流动着的不同液体有关。在此之后，这种说法还延续了1000年左右。其认为“甜”是一种积极的特质，鉴于蔗糖是已知的最甜物质，它被当成了一种灵丹妙药。除了其自身能抚慰伤痛，它还能与其他药物混合使用，让良药不再苦口；糖还能给食用它的人提供卡路里，也就是能量。

自9世纪起，威尼斯人就从埃及和东地中海地区进口糖蜜、糖和糖浆。蔗糖再从威尼斯出口到欧洲的其他地方。自13世纪起意大利北部的医学著作就在食谱和配方中提到了糖。例如，《健康全书》一书就根据阿拉伯11世纪的一份保健处方列举了糖的好处和坏处：

> 它能净化身体，对胸部、肾脏和膀胱都很有益处。危害：它会引起口渴，导致胆汁质的体液流动。中和之法：与酸石榴一起食用。功效：避免坏血的生成。它有益于所有气质、各个年龄段、每个季节和各地区的人。

威尼斯的药剂师专门提炼原糖，而且会熟练制作糖浆、果酱、坚果馅的糖果、紫罗兰糖和一种叫“青春神水”的东西，这种神水被他们吹捧为长生不老药。蔗糖还被当作礼物送人，一般是在结婚的场合，新娘会收到一盒甜食和一个用糖做成的婴儿小雕像。

蔗糖在14世纪的时候变得更常见了一些，也有更多的烹饪手册里提到蔗糖。在1300年一个版本的《肉食之书》（后来一个版本的作者被认为是纪尧姆·泰勒，人称“塔耶旺”）里，蔗糖只会出现在给病人的饭菜中。在1420年版本的相同菜谱中，糖被广泛地运用在大多数菜肴里。14世纪初的烹饪手稿《烹调之书》大概是在那不勒斯编纂而成的，书中的菜谱要求大量使用蜂蜜，但偶尔用蔗糖代替蜂蜜；在这本书中，蔗糖通常是和其他原材料混合使用，而非像蜂蜜那样被直接浇在做好的菜上。蔗糖仅被用来制作用香料调味的蚕豆、杏仁奶米饭、用糖和蜂蜜调味的蛋糕和用到苦橙的一些菜肴。然而到了14世纪，托斯卡纳的菜谱合集就非常注重蔗糖的使用了，蜂蜜只是被给予了一个边缘角色（大多是用在油炸馅饼和一些甜点里）。纵观阿尔贝托·卡帕蒂和马西莫·蒙塔纳里的《意式料理：一段文化史》（2003），书中的135个菜谱里，糖作为原材料的比例高达24%。

到15世纪，蔗糖已经在欧洲富人的家中十分普遍了；它被用来做酱料、点心和糖果。马埃斯特罗·马蒂诺1465年所著的《烹饪的艺术》中就大量提及了蔗糖。大概有50多个食谱要用到这种原料——卡仕达派、鱼和禽类料理、汤、蚕豆菜肴、糖蜜包裹的坚果、冷热饮品、锅煎芝士、油炸馅饼、焗通心粉和宽面条、蛋糕、杏仁蛋白软糖、牛奶冻和各式糖渍品。巴尔托洛梅奥·普拉蒂纳1474年出版的《合理享乐，身强体壮》一书中提及蔗糖的地方更多，而巴托洛米奥·斯卡皮在1570年出版的《大厨的艺术与技术》一书中，也出现了类似的情况。正如一则食谱所写的那样，蔗糖是“一切食物的最佳拍档”。

蔗糖一直被视为富裕的象征和财富的标志。当法国和波兰国王亨利三世 1574 年到访威尼斯城邦的时候，蔗糖是宴会上代表国王无上荣耀的重要组成部分。餐巾、桌布、盘子和刀叉——几乎桌子上的一切都是用糖制作的。在布景上也用到了 1250 个由雕塑家雅各布·圣索维诺设计的糖雕来吹捧王室，其中有个两虎相伴、骑在马背上的王后像，一侧胳膊上戴着法国国徽，另一侧则佩戴波兰国徽，其余的就是一些动物、植物、水果、国王、教徒和圣人的雕像了。

到了 17 世纪初，蔗糖在欧洲大陆已经很容易获得了，除了穷人还吃不到，大多数人的餐桌上都不难见到它的身影。自然而然地，糖代表的贵族威望也就被剥夺了。佛罗

来自利希的约翰·威廉姆婚宴上的糖制雕塑，克里弗和博格，杜塞尔多夫，1587 年。

伦萨人乔瓦尼·德尔·图尔科在《转瞬即逝的秘密》(1602)里就抱怨到，早期的食谱作家过于依赖香料和糖，这和“大多数人的口味不符”。用到糖的出版食谱数量开始下降，总体来看，在复杂的烹饪教程中糖的用量也在减少。

英国糖

在英格兰，亨利二世(1154—1189年在位)家族的账本上显示，当时他们使用了很少量的糖。关于莱斯特公爵夫人家庭的一项账目记载，在1265年的7个月时间里，他们只获取了55磅(约25千克)的糖，但这项开销的金额却不低。英国人，特别是权贵阶层，与糖的浪漫史始于14世纪。理查德二世写的专业书籍《烹饪之法》(约1390)就收录了很多用到糖的菜谱：比如油炸馅饼、卡仕达酱、派、酱料、炖菜、五香肉末、肉食、鱼、禽类、海鲜、野味料理和酒精饮料，包括塞浦路斯葡萄酒、德国葡萄酒和蜂蜜酒。在这些食谱中，糖会和多种多样的原材料一起混合使用，例如醋栗、鸡蛋、奶酪、葡萄干、枣、牛奶、杏仁露、无花果、梨、大米、面包以及几乎所有可得的香料和香草。有几个食谱还特别指明了塞浦路斯蔗糖或是其他特定的糖类。在那时，几乎是无糖不成宴。在约1438年写的《英国政策简论》一诗中，诗人就哀叹大量重要的商品都需要从佛罗伦萨和威尼斯进口，但唯独没有批评蔗糖：“如果有什么东西可以赦免于我的控诉之列，毫无疑问，唯有蔗糖一物而已。”《贵族烹饪指南》（约 1480）也包含了许多关于糖的菜谱，包括饮料配方，比如在红葡萄酒里加糖。大多数情况下，蔗糖的用量很少，菜谱的主要味道也不是甜味。

即使在很多宴会上，也没几道甜味的菜肴。这样的局面在 16 世纪前十年糖价下降之后得到了改善，即使不是很有钱的家庭也能负担得起蔗糖，它不再是贵族和皇室独享的奢侈品。诗人托马斯·纽伯瑞的咏叹调里就提到了糖果铺中卖的小面包（软包）、脆饼（烤硬的小蛋糕，掰碎的时候会咔嚓作响）、蜜渍物（果脯）和其他一些糖制品。糖价在这个世纪里持续下跌，但到了 16 世纪 90 年代，糖仍然是彰显地位的主要标志。在亨特福德爵士为伊丽莎白一世举行的宴会上，装饰性的食物被做成了平面的糖画和立体的糖雕大肆展示：三月千层糕、葡萄、牡蛎、肉、鸡冠花、长春花、蟹、龙虾、苹果、梨和各种李子。蜜饯、乳酪、果冻、果汁、果酱、糕点、糖果，各式各样。宴席上还有：

> 用糖做的各种城堡、堡垒、典籍、鼓手、小号手和士兵。狮子、独角兽、熊、马、骆驼、公牛、公羊、狗、老虎、大象、山羊、单峰骆驼、驴和各种野兽的糖塑像。鹰、隼、鹤、鸨、苍鹭、鹰、麻鸦、雉鸡、鹧鸪、鹌鹑、百灵鸟、麻雀、鸽子、公鸡、猫头鹰以及其他任何会飞的生物也都被做成了糖雕。还有蛇，蝰蛇、毒蛇，青蛙、蟾蜍，以及各种虫类。海里的美人鱼、鲸鱼、海豚、康吉鳗、鲟

鱼、梭子鱼、鲤鱼和鲷鱼。

到17世纪初，糖几乎在英国普遍受到赞扬。弗朗西斯·培根在他的乌托邦小说《新大西岛》（1624）里甚至建议，在纪念重要发明者的美术馆里为“糖的发明者”树立一座雕像。《克林克》（又称《节食之疾》，1633）的作者詹姆斯·哈特也曾说过，“蔗糖取代了蜂蜜，获得了更高的荣耀，也更易于入口；因此无论是为身体康健还是为疾病所累，到处都在频繁使用糖。”

在法国学习过的英国贵族的厨师格瓦赛·马卡姆在《英国主妇》（1615）一书中提供了几十种食谱。其中包括沙拉、烤肉、鱼、火鸡酱料以及其他的家禽、蜜饯、布丁、挞、甜味和咸味派、乱炖、蛋糕、煎饼、油炸馅饼、杏仁蛋白糖、果脯和很多别的菜式。《王后橱柜揭秘》（1655）的菜谱里也包含多种多样的糖类用法，比如蜜饯、蛋糕、奶酪蛋糕、煎饼、面包、糖花、南瓜和多样的馅饼、挞、布丁、豆类、果酱、沙拉调味品、酒精饮料（如牛奶甜酒和乳酒冻）、奶油和基础形状的硬糖做法，以及药用配方。

17世纪末的时候，对于英国的上流社会来说，糖已经失去了一些原有的魅力。罗伯特·梅的《成功的厨师》（1685）只提到了两则用糖的菜谱——分别是肉和鱼的酱料。约翰·伊夫林的《关于沙拉》（1699）中收录了十几个含糖的菜谱，但他明确表示，“除了需要用于调出更好的味道以外，糖几乎是被排除在料理界之外的，而它的料理伙伴其实有很多；比如一些酸味的调料，现在就很常用，虽然你执意不用的话也问题不大。”

在英国贵族对于蔗糖失去兴趣了之后，其他社会阶层的人逐渐发现了它的魅力，消费飙升。然而，真正让它重获新生的是饮料。

饮　糖

在中世纪时期，欧洲最受欢迎的饮品当属甜药酒或希波克拉斯酒（名字大概来自古罗马医学家希波克拉底），这是一种热过的加了糖或者香料的葡萄酒，常在饭后当作助消化的饮品。传统的做法是用蜂蜜来给希波克拉斯加点甜味。一位中世纪末期的医生阿纳尔杜斯·德·维拉诺瓦，他提供的法国希波克拉斯酒配方（约1310）里用的则是蔗糖。而中世纪另外一个用糖的食谱出自法国的烹饪指南《巴黎家事》（约1393），里面用了1.25磅（约0.57千克）的糖。在接下来的三个世纪里，希波克拉斯酒的配方时有出现。1692年的一个英国配方用了两夸脱[①]（约1.9升）的莱茵白葡萄酒、加纳利红酒和牛奶作为基底，再用1.5磅（约0.68千克）的糖调味。

希波克拉斯酒在18世纪就绝迹了，但在那时很多其他用糖调味的混合饮料大受欢迎。在英国和美国，比较受欢

① 夸脱是容量单位，一般有英制和美制两种用法。1英制夸脱约为1.1365升，1美制夸脱约为0.946升。

迎的就有蜜酒（用糖、糖蜜和蜂蜜调味的啤酒，有时候也会加朗姆酒来提高度数）和牛奶甜酒（用香料调味的热牛奶与麦芽酒或者啤酒调和在一起），最终发展成为蛋酒。乳酒冻——将用香料调味过的牛奶或者奶油连同甜酒或者苹果酒和砂糖一起搅打起沫——是节日场合用的烈酒。用柑橘榨取的甜柑橘汁、柠檬和青柠与各种酒混合就能得到人气很高的果汁甜酒；同样备受青睐的还有热甜酒，是用香料调味过的甜酒与朗姆酒和樱桃汁做成的樱桃浆混合而成。

卡莎萨，一种在巴西很流行的类似朗姆酒的酒精饮料。

夏天有冰潘趣酒，冬天也有对应的热潘趣酒。混合了红酒、糖和香料的桑格里饮料慢慢演变为桑格利亚汽酒。

在新大陆的欧洲殖民地，当地人会用糖及其副产品生产各种不同的酒精饮料。在葡萄牙殖民的巴西，有一种以糖为基底的卡莎萨酒，是一种烈性蒸馏酒精饮料。因为葡萄牙规模强大且颇有政治势力的白兰地酒产业会毫不留情地打击任何带来市场竞争的进口商品，因此巴西卡莎萨酒就留在了本地。然而，大概是巴西的荷兰和犹太移民后来把这种以糖为基底的酒精饮料理念及其制作方法介绍到了加勒比地区。在法属西印度群岛，它直接被称为“rhum”，这个词很可能源自巴巴多斯当地的英格兰人，有时也叫它“kill-devil”“rumme”和“rumbullion”。最终，这些名称被简称为“朗姆”。

糖蜜是制造蔗糖过程中的副产品，也会被用来制作度数低一点的酒精饮料。西印度群岛的奴隶会直接在糖蜜里加水，这样简单的做法就能让混合物发酵。18 世纪初的历史学家罗伯特·贝佛利记录了英国弗吉尼亚殖民地上最贫穷的人会用糖蜜做一种啤酒。这种饮品通常会用米麸、玉米、柿子、土豆、南瓜甚至耶路撒冷洋蓟来调味。在新英格兰，从西印度群岛进口来的糖蜜会用来做朗姆酒。新英格兰是个很适合做朗姆酒的地方：这里有足够的金属和技术工人来做蒸馏器，有很多货船可以将加勒比地区的糖蜜成桶运来，还有大量的木材来为蒸馏装置补充燃料以及制作酒桶。朗姆酒很快就成为美国人首选的酒精饮料。它可以直接饮用、

美属维尔京群岛中圣克鲁斯岛的克里斯琴斯特德朗姆酒蒸馏房中的景象，1941 年。

兑水喝，或者和像蔗糖这样的其他原材料混合饮用。其中最受欢迎的混合饮料就是潘趣酒，一般是用朗姆酒、柑橘汁和糖做的，具体的变化多到不计其数。牛奶潘趣酒要用到蛋黄、糖、朗姆酒和肉豆蔻，在派对和舞会上都是时髦之选。朗姆酒在英国本土的喜爱者也很多，但是在红酒业发达的欧洲大陆，商人纷纷谋求法律来限制这种进口酒饮。

但真正让糖在欧洲大陆，特别是在英格兰变得不可或缺的，是三种无酒精的饮品——巧克力、咖啡和茶。巧克力饮料发源于前哥伦布时期的墨西哥，当地的做法是把磨碎的可可豆用水冲开，再用香草、辣胡椒、胭脂树籽和其他原料调味。因为那时的新大陆还没有甜味剂，巧克力是一种非常苦的饮料。欧洲殖民者尝过之后，重新用香料给它调味，并且增加了甜度——最开始用蜂蜜，后来是蔗糖。

在16世纪初，巧克力和制作这种饮品的工具从中美洲引入了西班牙，但发展的速度非常缓慢。自西班牙开始，人们对于巧克力的喜爱蔓延到了意大利，进而扩散到整个欧洲。最开始，巧克力是用几味产自新大陆的原料调味，然后再用蜂蜜增加甜味之后饮用的。喝巧克力的风尚在贵族之间流传开来，蔗糖就逐渐代替了蜂蜜。欧洲最早的可可饮料配方可以追溯到1631年，出自西班牙医生安东尼奥·克尔梅内罗·德·莱德斯马撰写的第一本关于巧克力的专著：

> 取一百粒可可仁、两个辣椒或者线椒、一把茴香和两个香草荚——作为替代，还可以用六朵亚历山大玫瑰研磨成的粉末——两根肉桂棒，一打（12颗）杏仁和很多榛子、半磅白糖以及足以为饮品上色的胭脂树橙。如此，你就可以享用巧克力之王了。

后来，欧洲人又发现他们其实更中意没有异域香料调味的巧克力，但糖的使用还是保留了下来。热巧克力直到17世纪后半叶才成为英国重要的饮品。在伦敦，17世纪50年代有了巧克力专卖店，还有一些发行物大肆吹捧这种饮品的价值，并且提供制作的配方。在《伊甸园的亚当》（1657）中，威廉姆·科尔斯写道，“在伦敦的各个地方都能以合适的价格买到巧克力”，他还提到了其他的好处——巧克力是理想的春药。据他所说，巧克力有一种“十分有效的生育功效”。法国的第一个热巧克力配方出自弗朗索瓦·马西亚洛特的《皇室及资产阶级烹饪》(1693)。

咖啡饮品最早来自非洲东部和阿拉伯半岛。从9世纪开始传到了中东地区。游览过土耳其和阿拉伯国家的欧洲人将这种新奇的饮品记录下来，很多人都抱怨它过于苦涩的味道。一些埃及人通过加糖的方式企图“改变咖啡的苦涩”，至少曾在17世纪30年代到访过开罗的德国植物学家约翰·维斯令是这么说的。在17世纪中期，土耳其人在欧洲开了第一家咖啡馆。而威尼斯的第一家咖啡馆是在1629年开的，很快就在欧洲其他的主要城市开了很多分店。当欧洲刚刚有咖啡馆的时候，店里往往只提供黑咖啡，伴有可选添加的糖，没过多久，糖就成为咖啡离不开的伴侣。

几乎是在同一个时期，欧洲人爱上了巧克力和咖啡，茶叶也从东亚流入市场。中国人有几千年的饮茶历史，到了中世纪时期，茶叶通过丝绸之路的骆驼队走陆路运到了中东，之后到达俄罗斯。欧洲的探险家和那些游历世界的人在东亚尝到了茶，但直到荷兰人从中国进口茶叶（大概是1610年），这种饮品才被西欧世界所熟知。在17世纪中期，茶来到了英国，但最开始只是有钱人的享受。直到1658年，茶才出现在咖啡馆里，成了广受青睐的饮品。一位叫塞缪尔·佩皮斯的海军指挥官喜欢在日记中记录生活的小细节，他就记下了他是在1660年第一次喝到的茶。只用了短短几年时间，茶就和咖啡、巧克力、冰冻果子露一道入驻了伦敦大部分的咖啡馆，最后一种饮品是改良自中东的糖浆饮料。

有时候，咖啡、茶和巧克力中添加的糖分量很大。在1671年，巴黎的咖啡卖家菲利普·迪福尔出版了《咖啡、茶和巧克力的制作方法》，其中就描述了这几款饮品在“欧洲、亚洲、非洲和美洲”都是如何被饮用的。迪福尔也建议在咖啡中加点糖，但他也抱怨有的巴黎人做得太过了，他们的咖啡已经变成了“一杯黑色的糖水”。

英国的第一家咖啡馆是一位土耳其商人在1652年开的。这开始是件新奇事，后来成为一种风尚，然后风靡全城；到1675年，据报道，光是伦敦一座城市已经拥有超过3000家咖啡馆了，常客往往是贵族绅士和富商大贾。这些有钱人会一边啜着咖啡，一边讨论生意上的事和政治话题。

因为巧克力、咖啡、茶和糖不菲的价格，英国的咖啡馆一直是那些混得不错的人的地盘；底层人民则会聚在小酒馆里喝啤酒。后来，政府注资的垄断企业——英国东印度公司开始大量进口茶叶，年进口量从1725年的25万磅（约113000千克）飙升至1800年的2400万磅（约1900万千克）。随着货量的上升，茶叶的价格跌到了巧克力和咖啡之下。于是，茶消费量很快就超过了巧克力与咖啡的消费量，而且随着越来越多的茶叶进口，中产阶级也买得起了。茶成了英国人的首选热饮。

18世纪，英国不太富裕的家庭通常是将蜂蜜作为甜味剂，这很好解释：蜂蜜几乎比蔗糖便宜6~10倍。在这个时期，越来越多的蔗糖从加勒比地区的英国殖民地进口而来，随着价格的下跌，消费量顺势暴涨。在 18 世纪初，英国每年的人均糖消耗量是 4.4 磅（约 2 千克）。1784 年，茶的进口关税降低，随之而来的是茶叶使用量的急剧增加。人均食糖量涨了将近 600%，达到 24 磅（约 10.9 千克）。那些最穷的人，即使无力在餐桌上增添任何菜色，也能喝到加了糖的茶水。

一个重要的烹饪原料

在糖的价格落得比蜂蜜还低之后，它就开始不仅被当作饮品的甜味剂，还被用作烹饪配料。英国 18 世纪出版的烹饪书将糖引入了很多菜谱中。汉娜·格拉斯在 1760 年出版的《应有尽有的糖果铺》是英国第一本这种类型的书籍，书中大胆地将糖囊括在几乎所有食谱中，包括冰激凌、冰沙、奶油、果脯、蜜饯、柑橘酱、果酱、蛋糕、糖霜、面包、饼干、饮料、糖果、威化、杂烩、模制夹心烤肉饼、泡芙、挞的食谱，以及各种制作糖雕塑和糖渍水果、蔬菜、莓果、香料、坚果、种子、根茎和花的方法。伊丽莎白·拉法尔德所著的《经验十足的管家》（1769）里有 100 多种有糖的食谱：酱汁、馅料、馅饼、油煎饼、薄煎饼、燕麦粥、布丁、饺子、甜点、蛋奶沙司、糕点、棉花糖、雪浮岛①和各种饮

① 一种以蛋奶糊为“海水”，打发蛋白为“岛屿”的法式甜品。——译注

詹姆斯·吉尔雷，《回到凯尔希的大英雄》，又名《在圣詹姆斯的守卫日》，1797 年，图中的冰激凌尤为瞩目。

料——奶油葡萄酒、麦芽酒、各种葡萄酒、牛奶甜酒、冰冻果子露、果酒、白兰地和柠檬汽水。糖不再是奢侈品——它成为必不可少的食材。

正如糖在英国料理中占首要地位，它也在美国烹饪中占首要地位。虽然精制糖在美洲殖民地不算很便宜，但糖蜜的价格就非常低廉了，它可以作为甜味剂或者当作朗姆酒的基础原料。糖蜜还是饼干、蛋糕、派和布丁的主要甜味来源，也会用在玉米糊、蔬菜和肉食料理中，特别是猪肉料理。18 世纪 80 年代，一位英国的旅行者就曾抱怨，美国的每一餐都配着糖蜜，“即使是吃油腻的猪肉也要搭配它”。

糖以多种形式出售，最常见的是 8~10 磅（约 3.6~4.5 千克）重的圆锥形糖块。有钱人家会大量购入，但一个中产阶级家庭一年只会用一个锥形糖块。据估计，直到 1788 年，美国的年均食糖量也不过每人 5 磅（约 2.3 千克）左右。虽然世界范围内的食糖量在飞速增长，但人们与糖的爱情尚在萌芽阶段。糖的用量的确在饮料、肉类、派和蛋糕中增加了，但真正的暴增是在甜食和糖果里实现的。

[第五章]

甜食和糖果

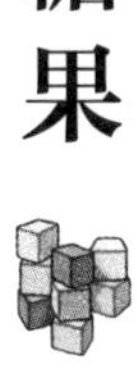

无论是在烹饪过程中腌沁入味，还是用糖衣包裹，食物被加入了足量的糖以后就不会变质了，因为糖分抑制了微生物的活动。这个特性使得商人在长途跋涉中，可以带着比如糖渍橙皮和糖衣包裹的杏仁这样的食物上路。固体的大糖块（比如冰糖和面包状的糖块）也能轻松交易。正是这样的贸易将甜食和糖果引入很多没有甘蔗生长的地区，或者一些不盛产特定水果和其他原料的地方。

虽然贸易路线的起点是南亚，糖果早在 7 世纪就来到了中东，后来又传到了欧洲。这些早期的糖果，例如果仁糖、果膏、杏仁蛋白糖、软糖和冰糖，是很多现代甜食和糖果的早期范本，它们流传至今，历史依旧有迹可循。只需简单列举一下就不难发现，很多欧洲早期的传统糖果都演化成了现代甜食——巴旦杏仁、果酱、甜派、蛋糕糖霜、软质太妃糖、硬质太妃糖、巧克力糖、硬糖球（很大块的硬糖）、柠檬糖、M&M 豆、甘草糖和冰激凌。

果仁糖（英文为“comfit”，源自法语“confit”，意为糖渍的。在意大利语里是“confetti”）最开始是有糖衣的药。医生和其他的药师用神奇的甜味物质包裹给每位病人配好的苦味种子、坚果、根茎、香料、香草和蔬菜提取物，无疑这可以让病人更从容地把药吞下去。病人还可能需要更多的卡路里，糖衣很容易提供；根据吃的糖衣药丸数量的不同，它们也可以给虚弱的病人带来一点能量的爆发。

果仁糖中间大多是有特殊香味的种子，比如茴香、香菜、丁香、葛缕子干籽或肉桂都很常见。在印度（还有其他地方的印度餐馆里），原味的或者有糖衣的茴香籽至今还会在餐后提供来帮助消化和清新口气。还有一种果仁糖是用从甘草根部提取出的物质制作而成的，甘草是一种小型豆科植物，原产自欧洲、亚洲和美洲。甘草根部有一股甜甜的茴香味，挤出汁并在锅中煮至黏稠之后可以做成一种糖果。这种叫“甘草糖”的糖果从中世纪开始就成了整个欧洲的一种重要甜食。

时至今日，世界各地还在生产形态、味道各异的甘草糖，虽然大多数用的已经不是甘草根提取物和茴香，而是人造香精了。在美国，最知名的甘草味糖果就是“Good&Plenty”牌的——嚼劲十足的小糖珠裹着粉色或白色的糖衣，令人想起传统的印第安果仁糖。这款糖果始于1893年。人工香精口味的扭绳甘草糖“Twizzler”是在1929年开始出售的。

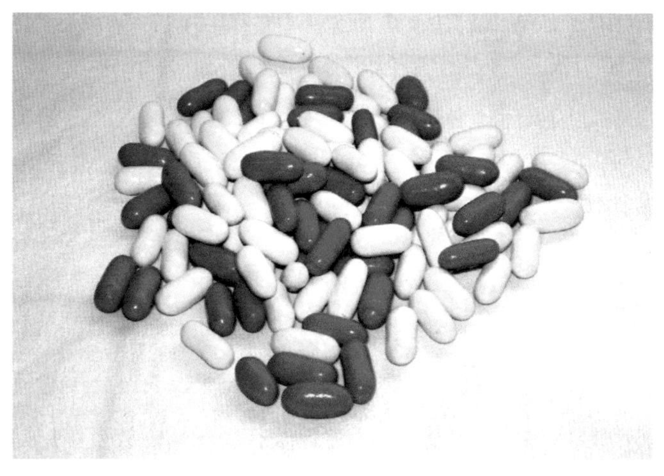

Good & Plenty 甘草糖，最早制造于1893年，时至今日还大受欢迎。

今天，大多数在美国销售的甘草糖都是用人工合成的原料大批量生产的，但是在其他地方，特别是在荷兰和斯堪的纳维亚，当地的国民零食还是真材实料的甘草糖，它们有各种形状，或软或硬，淡口甜味或者重口咸味。

另一种在中世纪时期比较常见的果仁糖就是有糖裹着的坚果，也发源于中东，之后被带到了欧洲。法语中的“dragées”一词就是专门形容用香料调味、糖衣包裹的坚果的，特别是糖衣杏仁。现在，这种食物在商业宣传中常常使用不同的名字，如巴旦杏仁，在中东也叫“mlabs”，而在希腊被称为“koufeta”。

糖渍水果和橙皮也是在中世纪的时候传到欧洲的，通常是在没有鲜果的季节里，在饭后食用的。它们到现在还以不同的形态存在——蜜饯或果脯、巧克力脆皮樱桃、果酱、柑橘酱和果冻。

杏仁蛋白糖（在英语里是“marchpane”）是一种用杏仁粉和糖混合而成的厚重糖膏，在中世纪时是中东地区大受欢迎的美味。它有可能起源于伊朗，后来通过阿拉伯文化的影响传到了欧洲。在欧洲，对它最早的记述是在13世纪晚期的意大利北部。但在此之前，用蜂蜜制作而成的杏仁糖很可能已经在西班牙、加泰罗尼亚和意大利广为流行了。杏仁蛋白糖在法国、德国、荷兰、北欧和英国也都备受青睐。它不仅仅是一种一口大小的糖果，它还会被塑成小猪或者鸡蛋的形状，而且经常会在特殊场合被当作礼物赠送，例如圣诞节、复活节或者婚礼上。杏仁蛋白糖在整

巴塞罗那波盖利亚市场上的糖渍水果

个欧洲以及许多欧洲先前的殖民地仍然很受欢迎。

拉丝糖也是在中世纪由中东地区的阿拉伯人传到欧洲的。它的制作方法是先把糖用水溶化，然后揉捏成可塑的状态，可以被拉伸并拉成不同的形状，比如缎带、花朵或叶子。

在中世纪，糖也被当作餐后的消化剂——在正餐被撤掉以后，会端上加了糖和用香料调过味的葡萄酒作为一餐的结束。这道食物被称为餐后甜点（“dessert”来源于法语中的“desservir”一词，是清理餐桌的意思）。到了18世纪，甜点已经成为一道精美的餐点，可能会出现奶油、果冻、挞、派和甜布丁。虽然通常是在晚餐后食用，这种甜食也可以在下午或者傍晚食用。制作甜点成为一门家务事的艺术，通常是由一位专业的甜品师或是一位用人在专业的指导下完成。

巧克力糖最初是在17世纪的法国被发明出来的。

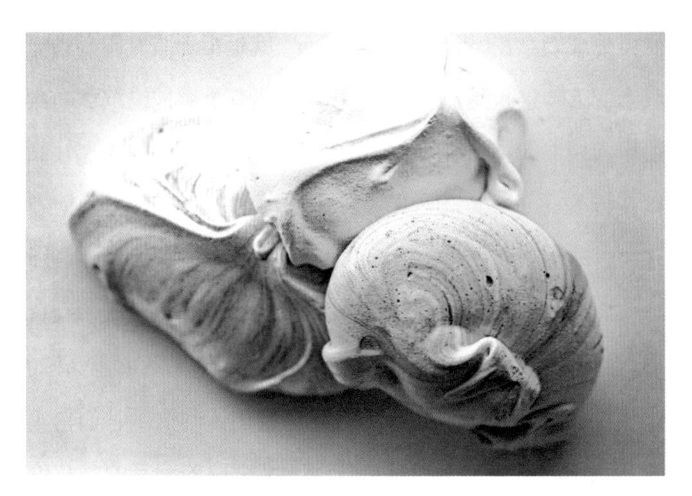

自 18 世纪以来，蛋白霜的食谱就已经出版了。

劳拉·梅森在她《糖李子和冰冻果子露》（1998）一书中写道，到了 18 世纪，在英国的糖果铺里已经可以买到蜜饯和糖渍水果、饼干、蛋糕、马卡龙、糖蜜、果仁糖、派、挞和各种装饰用的糖雕了。糖果铺还销售从其他国家进口的糖果。巧克力糖（各种样子的华丽糖果，早期专供法国宫廷）被进口到英国和其他欧洲国家。这些糖果通常是裹着巧克力的软糖、水果和坚果，是有闲阶级——唯一买得起它们的人——的奢侈品，但这样的局面很快就会发生变化。

甜品师

到了 17 世纪，做果仁糖的人和甜品师会在小店里卖他们做好的甜食供人们回家享用。一段时间之后，零售商发展得更成熟。在巴黎，“limonadier”就是指那些专门卖柠檬汽水的人（这也是他们名字的由来）。有个卖柠檬汽水的西西里人叫弗朗切斯科·普罗科皮奥·代·科尔泰利，他1686年在巴黎开了一家咖啡馆。除了提供咖啡，店里还卖糖衣水果、冰冻甜食、甜味的冷饮、酒精饮料和热巧克力。

18世纪和19世纪期间持续下跌的糖价让甜食成了大众买得起的食物。在英格兰，从18世纪80年代到19世纪20年代期间，乡镇城市的甜品店数量增加了四倍。它们售卖五花八门的糖果，有的是从其他国家进口的，有的是从伦敦寄过来的。

特殊的糖果种类也在这时诞生了。冰糖是将蔗糖和水

冰糖的历史大概有2000年之久；时至今日，它的吃法还很多，其中一种就是凝结在一根小棍上吃。

按照过饱和的状态混合，一段时间之后析出的晶体；而硬糖则是在煮沸糖水得到糖蜜之后，把它倒进模具里或者直接用手塑形。这些逐渐演变为大块硬糖、棒棒糖、薄荷糖和拐杖糖果。糖衣坚果是坚果脆糖的母版——后者是将整颗坚果仁或坚果碎嵌入一块黄油质地般的糖浆上得来的硬糖。水果硬糖最初是用真正的果汁给煮沸的糖浆调味后制成的，在很早的时候就大受欢迎，但今天大多都是用玉米糖浆、人工调味剂和色素来仿制水果的味道和外形。“救生员”是美国和加拿大最有名的现代水果糖品牌。

软质与硬质太妃糖

软质与硬质太妃糖最早出现在19世纪早期的北英格兰，在那里，大多数糖果都是在家或者糖果铺中制作而成的。（埃弗顿的糖果工匠制作的太妃糖是远近闻名的，那里现在被划为了利物浦的一个区。）基础食谱要用到糖或者糖浆，连同黄油和调味品一起煮沸，比如橘子、柠檬、巧克力或者香草这些调味品。基于这个基础配方，形成了两种截然不同的甜食。

制作软质太妃糖要在混合物煮制“硬球”阶段——也就是当糖浆滴进冷水能够迅速形成坚硬的小球的时候——迅速离火。待糖浆稍微冷却，借助一个金属钩用手拉伸它，把糖拉到如绸缎般丝滑才可以。在19世纪，“拉太妃糖”在派对上很受欢迎，宾客们会成双成对地用涂满黄油的双手拉伸太妃糖，接着享用自己的劳动成果。

做硬质太妃糖则需要把混合物加热到“硬裂纹”阶段，也就是说，糖浆滴到冷水里会形成易碎的脆丝线。冷却下来以后，糖浆就能形成浓缩糖果，糖果碎了会发出啪的一声。在英国，硬质太妃糖常常是盖伊·福克斯之夜（每年的 11 月 5 日，也叫“篝火之夜”）的代名词，会以“篝火太妃糖”的名称贩卖。

软质和硬质太妃糖从英国一路漂洋过海，在 19 世纪 40 年代来到了美国，两者在东海岸的城市，特别是在费城和大西洋城都备受推崇。来自宾夕法尼亚的约翰·罗斯·埃德米斯顿可能是第一个卖“盐水太妃糖”的人，大概是因为

约瑟夫·威廉·桑顿 1911 年 10 月在谢菲尔德开了他的第一间糖果铺；他店里最受欢迎的产品就是硬质太妃糖。

有一次暴风雨把海水卷到他位于大西洋城的糖果铺里，他才发明出来这个产品。事实上，普通的太妃糖和盐水太妃糖在制作方法上没什么差别，但光是名字就让后者显得更有吸引力一些。其他人又改良了配方，拓展了产品线，让它有了很多吸引人的炫彩颜色、多样的口味和不同的形状。到了20世纪20年代，美国有超过450家公司制作盐水太妃糖，很多都是在临海的度假胜地。

制造商

在欧洲和北美，手工制作的糖果从18世纪末就开始作为商品出售，但一直到19世纪，糖价下跌、科学技术让蔗糖提纯变得更加便捷之后，糖果才被广泛食用。糖果的首次批量化生产是在19世纪50年代的英国实现的，但这样的制造技术很快就在其他国家普及开来。糖果以前所未有的规模大批量地出现在人们的生活里，随着时间的推进，有了更多的形状和大小的糖。到了19世纪末，中东、欧洲和北美地区已经建起了上百家商用的糖果制造厂。他们生产的大多数是小硬糖，通常在商店零售，被摆在大玻璃罐里以几分钱的价格出售。

又软又有嚼劲的糖也会被大批量生产。其中一个源自中东的广受欢迎的产品就是土耳其软糖，也叫“rahat loukoum”（翻译过来是“喉咙的休息时光”）。它是用糖（最开始是蜂蜜）、淀粉或者像阿拉伯树胶这样的胶状物质和调味品制成的。玫瑰水或者橙花水都是比较常见的调味品选择，像杏仁、开心果、榛子等坚果碎或者切片的果干也可以加进去。做好的混合物要在一个铁盘中放凉，然后切成方块，最后撒上（粉状的）糖霜。这个发明要归功于一位18世纪中叶的土耳其糖果商。这种糖果迅速在中东和欧洲流行起来，特别是在英国，从1914年开始，英国就有了土耳其软糖巧克力棒，这是一种里面有玫瑰味内馅、外面有牛奶巧克力脆皮的软糖。

软糖豆——小颗的、豆子形状的糖果，有着密实的果冻胶内心和一层硬糖衣——有可能是土耳其软糖后来的衍生品。它有令人眼花缭乱的颜色，对应着不同的水果口味。关于软糖豆最早的纸质记录是1886年一则伊利诺伊州的广

软糖豆在圣诞节和复活节都是广受青睐的糖果。

告，它被当作圣诞节的糖果销售。软糖豆通常是放在糖果铺的大罐子里出售，或者是在自动贩卖机里。到 20 世纪 30 年代，软糖豆又被营销为复活节的专属糖果，大概是因为它们的形状有点像鸡蛋。

1869 年，阿尔贝特和古斯塔夫·格利茨这两位德国移民在伊利诺伊州的贝尔维尔开了一家糖果铺，他们几乎可以称得上是当代软糖豆之父。到了世纪之交的时候，他们的家族生意专注于制作奶油糖果，比如玉米糖，一种形似放大的玉米粒的三色糖（黄、白、橙三色）。1976 年，格利茨的后代开始尝试“美食家”软糖豆，这是一种比正常大小小一些的糖果，以一些意想不到的味道为特色，例如梨子、西瓜、根汁汽水和黄油爆米花（据说这是最受欢迎的一款）。他们给这个新产品取名为“Jelly Bellies”，这个品牌有 50 多种口味，比如卡布奇诺、辣芒果和椰林飘香鸡尾酒。现在，这个公司生产“比比多味豆”，这个名字来源于 J. K. 罗琳《哈利·波特》里面提到的一种产品；此外，他们还推出了富含维生素 C 和电解质的运动能量豆。

有嚼劲的糖果在中世纪就已经出现了，是在中东制作的。在这类糖果中，最早进行商业广告推销的是蜜枣糖，取这个名字正是因为它的主要原料是枣胶（来自一种枣属的灌木）。现在，这种果味的糖粒一般都是用土豆淀粉、树胶和糖或者其他甜味剂制作的。而另一种有嚼劲的糖果紧随其后出现，那就是水果或者蔬菜形状的果汁软糖。小熊软糖是在 20 世纪 20 年代被德国人发明出来的。这些五彩斑斓的小造型的基本成分就是动物性的吉利丁。1982年，德国糖果公司哈里波率先在美国制造出了“橡皮”糖。另一家德国制造商口力在20世纪80年代将虫形橡皮糖引入了市场，此后一直都很受欢迎。瑞典小鱼软糖是另一款火爆的橡皮糖，是在20世纪60年代从瑞典进口的、不添加动物吉利丁的产品。纵观世界，橡皮糖以千姿百态的造型和变化万千的味道彰显着自己独特的魅力。

节日甜食

很多甜食和糖果都是与节日息息相关的，特别是圣诞节、光明节、复活节、万圣节和情人节。在糖还是昂贵的珍品时，不太富裕的人家就只能在这些特殊日子里才能吃上点甜的。

圣诞节水果蛋糕的传统可以追溯到中世纪。最初，这种蛋糕是通过糖渍水果增加甜味的，具体做法就是把糖渍水果搅入面糊或者面团中；到了16世纪，糖已经是很普遍的原料了，糖霜也变成了普遍的配料。不同的地方有各自独特的习俗，比如英国的圣诞节蛋糕和德国的史多伦蛋糕。

那些要庆祝12天圣诞节（12月25日至1月6日）的文化，往往都有各自的传统蛋糕。第十二夜，或称主显节，也叫三王节；在法国，当晚要吃galette des roix（国王蛋糕）。在西班牙和拉美地区，传统的糕点是环形的三王节面包（也叫“国王的戒指”），上面会慷慨地铺满糖渍水果。很多国家都会在主显节这天吃各式各样的“国王蛋糕”；在一些地方，特殊的国王蛋糕也是“肥美星期二”[①]的专属食物。

在不同的历史时期和地域文化中，圣诞节上受欢迎的甜食种类繁多，比如咸奶油糖果、巧克力、柠檬、奶油、焦糖、软糖豆和其他甜的东西。一直到19世纪中期，拐杖糖——顶端带一个弯钩的红白条纹硬糖棍——才成为美国节日庆典的一部分。拐杖糖的发明要归功于俄亥俄州伍斯特的奥古斯特·伊姆加德，据说他用装饰纸艺和拐杖糖制作并装饰了圣诞树。它并非一问世就大获成功，虽然有一小部分家庭会在家自己制作拐杖糖。这种糖不仅做法很复杂，还很容易碎，所以也比较难运输。但这一切都在20世纪50年代得到了改善，那时有了可以自动制作拐杖糖的机器，而且包装工艺上的发明创造也帮助拐杖糖得以在到达目的地之前保持完好无损的样子。像玛氏、好时和雀巢这些糖果巨头现在都有他们自己品牌旗下的拐杖糖。

万圣节（或者万圣节前夜，就是万圣节的前一天晚上）在说英语的国家往往是定在10月31日的晚上，小孩子们会穿上特殊服装挨家挨户要糖果。因为万圣节正好是很多国家苹果丰收的季节，经典的庆典食品就有糖渍苹果、太妃糖苹果或者焦糖苹果。像糖爆米花球和软质太妃糖这样

① 大斋期前一天的忏悔日。——译注

在家中制作的甜食逐渐让位于外面买的糖果，特别是19世纪80年代发明的玉米糖（三色的、玉米粒形状的糖果）。现在，甚至能买到小盒装的或者独立包装的单个糖果，专门供万圣节的时候发给小孩子。时至今日，这个满是糖果的美国版万圣节也被很多其他的国家学了过来。

在非基督教徒的春日庆典上，丰饶是一个频繁使用的意象——兔子、蛋和鸡就这样被吸纳进了基督教的复活节庆典中。从中世纪开始，将复活节蛋赠予贫穷的孩子就是欧洲复活节的传统之一。而复活节糖果则是一个相对较晚出现的传统，大概起源于东欧。第一个有记载的巧克力复活节蛋出现在1820年的意大利。在20世纪30年代，像软糖豆和巧克力兔子这样的复活节甜食成为复活节篮子传统的一部分。在1953年，美国的新生糖果公司开始制作3D的复活节小鸡棉花糖，叫作“Peeps”。2012年，美国人一共在复活节糖果上花掉了超过23亿美元，其中包括9000万个巧克力兔子、7亿个棉花糖小鸡和160亿个软糖豆。

为期八天的光明节是犹太教庆祝军队在公元前164年击败塞琉古帝国、重建耶路撒冷第二寺庙的日子。光明节往往是在家过的，每晚会给孩子送上一个不太奢侈的小礼物。礼物通常是小硬币或者一点钱。在20世纪20年代，城里的糖果制造商开始宣传他们的产品是理想的光明节礼物。比如纽约的阁楼糖果公司就会卖金箔纸包装的圆形扁平巧克力，以模拟硬币的形状。总部在布鲁克林的巴顿公司成立于1938年，会为光明节和逾越节准备特殊的犹太洁食巧克力。

情人节（2 月 14 日）据说是为了纪念古罗马时期一位被杀害的圣徒，在欧洲中世纪就是一个流行的节日。糖果具体是在什么时候变成了情人节传统的一部分，我们尚不清楚，但在 1860 年，英国糖果商理查德 · 卡德伯里设计了第一个情人节巧克力礼盒。时至今日，情侣还会在这天交换精美的巧克力礼盒。在美国，新英格兰糖果公司于 1902 年发明出甜心糖果，是一种小巧的心形糖果，上面刻着一些简短的浪漫箴言。到 21 世纪，这家公司每年大概会生产 80 亿个甜心糖果——几乎全部的货品都会在情人节前六周被抢购一空。

巧克力

在欧洲和北美，17 世纪中期就有加糖饮用热巧克力的传统了，但直到 19 世纪巧克力才被当成一种甜食贩卖。1815 年，荷兰人昆拉德 · 范 · 豪滕发明了一种给巧克力脱脂、碱化的工艺。这一系列的研究最终让人们发现了制作可可粉的诀窍。终于，这种产品在 1828 年横空出世。也正是这个壮举，让大规模生产巧克力粉和固态巧克力成为可能。

19 世纪中叶，英国开始生产手工制造的巧克力。一位巧克力制造商约翰 · 吉百利，亦是一位贵格会教徒以及强烈提倡戒酒者，他认为提供酒的替代品非常重要。1831 年，他的吉百利公司开始生产用于制作巧克力饮品的可可；到1866年，吉百利公司还生产巧克力食品——比如手作巧克力夹心糖、巧克力脆皮牛轧糖和其他的巧克力糖果。从1897年开始，吉百利公司开始生产牛奶巧克力。另外一个影响颇深的巧克力制造商是约瑟夫·斯托尔斯·弗赖伊，他也是一名英国的贵格会教徒，发明了将可可粉、糖、融化的可可脂混合而成的巧克力糊，可以在模具里塑形制作巧克力棒的方法。很快，J. S. 弗赖伊家族公司就成为全世界最大的巧克力制造商。然而，在1919年，吉百利公司收购了J. S. 弗赖伊家族公司，吉百利公司生产的一些巧克力棒到现在还冠着“弗赖伊的”名号。

而吉百利公司又在2010年被卡夫食品公司收购了，但仅仅三年之后，卡夫就把旗下的甜品和零食制造产业链分给了一个新公司——亿滋国际集团。现在，亿滋最畅销的糖果品牌分别是吉百利牛奶巧克力、米尔卡巧克力和无糖口香糖，分别占据全球销量排行榜的第六、第五和第三。

英国约克的零售商威廉姆·图克父子在1785年就开始卖可可。1862年，亨利·艾萨克·朗特里收购了图克的可可业务。在此之后，公司在1881年引进了朗特里的果味软糖，又在1893年推出了果味口香糖。四年之后，朗特里有限公司成立了。就像吉百利公司，朗特里公司为员工提供了非常优渥的福利，例如员工餐和其他设施、工会、养老金计划、失业补贴和带薪年假。

1931年，朗特里开始了雄心勃勃的扩张计划。而公司

店主卢德女士正在图克的店里向青少年卖巧克力，当时正好是 1926 年 1 月 2 日的学生展览会，地点在伦敦威斯敏斯特的园艺厅。

成功的关键正是与福利斯特·马尔斯非同一般的关系，后者在1932年的时候在英国推出了玛氏糖果棒。在那之前，“混合糖果棒”（其中包括巧克力、花生、焦糖等多种原料）在英国的名气还没那么大。1935 年，朗特里公司推出了“巧克力脆”这个产品，两年之后改名为“奇巧”。朗特里公司在 1937 年推出了“聪明豆”。这种五颜六色的糖衣巧克力豆至今在英国、南非、加拿大和澳大利亚都还很受欢迎。1988 年，雀巢收购了朗特里公司，在此之后一直用这个公司的品牌推出新的巧克力和糖果产品。

美国的巧克力制造商

米尔顿·赫尔希本是宾夕法尼亚州兰开斯特的一位焦糖制造商。1893 年时，他参加了在芝加哥举办的哥伦比亚博览会，展览中德国德累斯顿的莱曼有限公司设计的巧克力制造机给他留下了难以磨灭的印象。在博览会上，赫尔希就直接买下了这套设备，并把它运回了兰开斯特。接着他从贝克巧克力坊雇用了两名技师，开始了批量化的巧克力生产。在那个时候，美国所有的巧克力都还是手工制作的。作为好时焦糖生意的一个副线，好时巧克力公司在开始的时候主要生产早餐可可、甜巧克力、烘焙巧克力和其他小糖果，终于在 1905 年推出了最著名的好时牛奶巧克力棒以及 1907 年的好时 Kisses 巧克力。诚然，好时巧克力公司一开始就顺风顺水，它在一战期间又受到了极大的推动。那时，好时的巧克力棒会被供给欧洲战场上的美国士兵。很多人在此之前从来都没吃过巧克力棒，在他们从战场返回家乡之后，对于好时巧克力的需求迅速飙升。近来，好时公司一直通过提高销量和兼并来获取更高的国际市场份额。从销量上看，它旗下的瑞兹巧克力是在美国卖得最好的产品，在全球销量排名第四。

1992 年，明尼阿波利斯市的弗兰克·马尔斯创建了一家叫作“Mar-O-Bar”的公司，最开始是卖含有焦糖、坚果和巧克力的糖果棒。一年之后，公司推出了“银河巧克力棒”，并在 1930 年推出了“士力架”。这是一种花生味的牛轧糖糖果棒，上面撒满了坚果，外面还有巧克力脆皮。它很快就成为美国最受欢迎的糖果棒，自此从未被超越。

弗兰克·马尔斯和他儿子福利斯特·马尔斯的关系并不十分融洽，后来，他儿子就拿了5万美金，只身一人前往

奇巧威化巧克力是英国人发明的；现在，它几乎可以在世界上的任何一个角落找到。

瑞兹巧克力片出现在了商业大片《ET》中。

福利斯特·马尔斯在 1932 年时将玛氏糖果棒推广至英国。

英国创立了一个新的公司——玛氏有限公司。1932 年，他推出了玛氏糖果棒，这是一个比银河巧克力棒甜度低一些的版本。到了 1939 年，玛氏已经成为英国第三大糖果制造商了。在 1939 年第二次世界大战爆发之后，福利斯特·马尔斯回到了美国，和布鲁斯·默里合开了一家新公司，后者是好时公司总经理的儿子。因为他们两人的姓氏都是以“M”开头的，他们就给新公司命名为“M&M”。他们的第一个产品是小的牛奶巧克力豆，外面有硬质的彩色糖衣，很像一个小版的士力架，他们将这个玩意取名叫 M&M 巧克力糖。一直到 1964 年，玛氏和 M&M 才合并成为一家公司。玛氏一直通过兼并和推出新产品扩大自己的影响力。2011 年，玛氏已经掌握了 15% 的全球糖果市场，是世界上最大的糖果制造商。他们的德芙巧克力（在英国叫银河）、奥碧

玛氏的银河巧克力棒是第一批火起来的混合口味糖果棒。

口香糖和益达口香糖在世界上的销量排名分别是第五、第八和第九。几十年来，玛氏和 M&M 一直都是世界上数一数二的糖果公司。M&M 的营业额在 2012 年达到了 34.9 亿美元，但不敌玛氏的士力架，后者在世界范围内的销售额高达 35.7 亿美元，是现在全世界卖得最多的糖果产品。

其他糖果制造商

在 19 世纪 60 年代，住在瑞士的德裔药剂师亨利 · 雀巢发明出了炼乳，并且成功地在市场上推销了一个以牛奶和面粉为主的婴儿食品配方。在 1874 年，他卖掉了自己的公司，但买家没有更换这家公司的名字。亨利 · 雀巢在他朋友、巧克力师丹尼尔 · 彼得的帮助下改良了牛奶巧克力棒的配方。彼得的巧克力加上雀巢炼乳让他们的产品一跃成为欧洲最知名的巧克力品牌。

阿华田是 1904 年一个由医生研发的瑞士产品，最先用来给严重的病患提供营养，在之后的几十年里，逐渐在市场范围内推广。它是一种加了糖和麦芽糊精的巧克力粉，可以用来冲奶，当冷饮或热饮都可以，添加了维生素的保健功效是它广告中吹嘘的重点。阿华田的成功让雀巢也开始推销自己的牛奶冲剂——“雀巢巧伴伴”，一种甜味的巧克力速溶粉。它是在 1948 年推出的，长期赞助儿童电视节目的策略也保障了它长久的人气。后来，这条产品线又继续拓展，引入了糖浆和麦片制品。

一战抑制了雀巢的销量，但在战争结束之后，公司继续迈出扩张的步伐。到 20 世纪 20 年代末，巧克力已经成

美国几种最热销的糖果

位于美国威斯康星州马尼托瓦克的比尔特森糖果铺，是一家很有年代感的老式糖果店。

巧克力棒的制作和销售遍及发达国家。

为该公司第二重要的产品了。1929 年,雀巢收购了丹尼尔·彼得的公司，开始生产做巧克力奶、优质巧克力和固体巧克力棒的巧克力粉。二战结束之后，公司飞速发展，部分原因在于收购别的公司。1988 年，雀巢陆续收购了意大利巧克力制造商佩鲁吉娜和英国的巧克力制造商朗特里。它旗下的奇巧威化在全球销量榜上名列第十。

另一家大型的糖果企业是不凡帝范梅勒集团，它是在 2001 年成立的，总部在意大利的米兰。它旗下的曼妥思薄荷软糖在今天的全球销量榜上排在第十一位。

现在，全世界遍布着数以万计的各式各样的糖果品牌。瑞典人是全世界糖果食用量最高的人群（每年每人 37 磅，合 16.8 千克）。而瑞士人是全世界巧克力食用量最高的人群（每年每人 25 磅，合 11.3 千克）。美国人均食糖量虽然没他们多，却支付着最高的金额——每年 320 亿美元，而且这个金额还在持续上涨，即使在经济萧条的时候也是这样。世界范围内的糖果销售量也在上涨，现在来看，年销售额在 1500 亿美元左右。

[第六章]

美国福音

除了糖果和甜食，糖在全世界大量的食品中都以添加剂的形式存在，例如早餐麦片、饼干（曲奇）、甜甜圈、冰激凌和软饮。即使不甜的速食品也往往含有糖分，只是它不算主要的调味品而已。这种隐形的糖就存在于汤品罐头和蔬菜罐头中；当然还少不了面包、薄脆饼干和薯片，速冻餐食、调味品（番茄酱、辣椒酱和伍斯特沙司）和沙拉酱、花生酱、婴儿食品和婴儿配方奶粉、比萨、热狗和午餐肉，泡菜和鸡尾酒小吃，风味酸奶、冷冻食品、果汁、水果冷饮、能量和运动饮料，甚至宠物食品。糖顶着各种名号藏在这些食品中，比如蔗糖、葡萄糖、右旋糖、麦芽糖、乳糖、半乳糖、麦芽糖浆、麦芽糊精、玉米糖浆、高果糖玉米糖浆、糖蜜和玉米甜味剂。

世界上没有哪个国家比美国更常用糖制造加工食品。随着糖价在 19 世纪下跌，无论挣多少工资、位于哪个阶级，甜点和零食在美国家庭变得普及。越来越多的糖用于制造蛋糕、饼干、派和其他糕点。甚至连外国游客都注意到了这件事，他们觉得“无论富有还是贫穷的美国家庭，用糖和其他甜味剂的分量简直大到惊人”。到了 19 世纪 70 年代，美国每人每年的食糖量大约是 41 磅（约 18.6 千克）；随着商业化的食品生产投入市场，蔗糖价格进一步下跌，美国人的食糖量随之增加。无论是简单还是繁复的蛋糕都成为美国人日常饮食的一部分。聚会庆祝时会摆上足量的果冻蛋糕、磅蛋糕、梅子蛋糕和女士蛋糕。美国的厨房里源源不断地飘出持久而馥郁的香气，那是糖饼干、华夫饼、好时巧克力、糖豆、环形小甜饼、坚果脆、马卡龙、姜饼、油炸圆饼和甜甜圈的味道——其中的含糖量都越来越多。甜的面包卷和甜甜圈成了常规的早餐食品。到 1901 年，美国人每年的平均食糖量已经达到了 61 磅（约 27.7 千克）。美国人对于糖的热爱正在不断升温，但是距离达到巅峰还有很长一段路。

棉花糖是一种在纸卷上拉丝形成的膨化糖，在 20 世纪初的美国十分受欢迎。

早餐麦片

一直到 20 世纪，经典的美式早餐还包括水果、面包、鸡蛋、土豆和各种肉类——不仅仅是培根或者香肠那么简单，甚至会有牛排、咸味肉派和小牛肝。在 19 世纪末，素食主义者和注重健康的改革派开始以未精加工的糙米为基础发明早餐食物，他们认为这种粗粮更适合现代白领工作者的消化系统。最开始，麦片这种商品是不甜的，而且要用白水泡着吃。随着这个产业越做越大，制造商发现顾客更喜欢吃甜麦片，于是就开始宣传吃麦片搭配奶油和糖这种新方式。威尔·凯洛格在玉米脆片的配方里加入了糖，为此还和哥哥约翰·哈维·凯洛格产生了严重分歧，后者是一位健康食品权威和素食主义者，他认为摄入糖比吃肉食给身体带来的潜在健康风险还要大。

20世纪，越来越多的妇女涌入了劳务市场，广告就将麦片标榜为减轻母亲工作负担的好办法。小朋友可以独立完成早餐的制作，而且他们也都很喜欢麦片甜甜的味道。据当时的医学专家所说，麦片对小孩的身体很好，对于繁忙的母亲来说是双赢的选择。大概是因为主打儿童市场，这些麦片公司即使是在大萧条的时候也赚得盆满钵满；不断提高产品的含糖量成为商家达成交易的秘密武器。二战之后，糖配给制成了过去式，麦片制造商进一步加大了赌注。1949年的时候，波斯特麦片推出了“糖脆”，是一种裹着脆糖衣的膨化小麦片。这个产品一炮而红，其他麦片厂商也纷纷效仿，生产出针对儿童市场的重糖麦片。有些谷物片里面一半都是糖；家乐氏的蜂蜜脆里，糖分占了55.6%；对手波斯特立刻出了有70%含糖量的超级橘子脆。这不禁让观察者发问：“这到底是麦片还是糖果呢？”

这些高糖的谷物片在儿童频道上铺天盖地地宣传，特别是在收音机和电视里，还有零售店的营销活动，以及后来的网络。美国的麦片三巨头（家乐氏、桂格和波斯特）花在广告上的钱比他们用来购买原料的成本还要高。每年，美国的麦片制造业要用掉 8.16 亿磅的糖，每份大概有 3 磅（约 1.4 千克）。讽刺的是，冷麦片早餐最开始是作为健康食物推出的，现在已经成为美国高糖饮食最大的元凶之一，特别是对小孩而言。每年，美国电视上会播出 1300 多万条麦片广告，多数都是针对儿童观众的。

美国超市中有很多各式各样的早餐麦片。

饼干、曲奇、蛋糕和面包

英语单词“biscuit”经由中古法语源自拉丁语，本来是“二次烘焙”的意思。一些欧洲早期的食谱（比如那些在意大利仍被称为“biscotti”的食物）是先将面团以条形烤制，然后分成或切成小片，再烤制一次，小心翼翼地去除多余的水分；成品越干燥，能够保存的时间也就越长。英国饼干的食谱随殖民者一起来到了美洲大陆；但美洲的一部分是被荷兰殖民的，当地的“koekje”一词本来是小蛋糕的意思，后来成为美国人对甜饼干的称呼。《美国烹饪》（1796）的作者阿梅莉亚·西蒙斯出版了第一则有迹可循的“曲奇”食谱，还包括一个用一磅半糖和三磅面粉混合而成的圣诞曲奇食谱。

曲奇是最简单的烘焙食品，只需要用到很少的原料——其中之一就是糖。曲奇只需要在烤箱里待上短短十几分钟，不怎么费神就能烤好，所以经常作为非正式甜点或者零食来享用。满满当当的曲奇罐常常被视为一个理想美国家庭的标志，代表了慈祥的母亲和幸福的家庭，当然，罐子里的曲奇一定要是自家制作的（这点甚至都不需要特意说明）：比如刻成不同形状的曲奇、燕麦曲奇、花生黄油曲奇或者经典的巧克力豆曲奇。隔壁的烘焙坊或者糕点店往往会提供更精致的款式，美国的商业化烘焙工厂从19世纪起也开始大量生产曲奇饼干。到了20世纪初，全国都能在商店买到曲奇饼干了；这些厂家还进行广告营销，企图赢得追求时髦的买主的芳心。美国国家饼干公司（后称纳贝斯克）创立自1898年，是一个由小型烘焙公司组成的集团，开创了各种各样的饼干，比如现在世界上销量最大的奥利奥饼干。

圣诞节饼干（曲奇）从16世纪就出现在欧洲了；时至今日，它还是圣诞节必不可少的食品。

正如我们今天所知，蛋糕是由面包演变而来的。有些像松饼一样，是扁扁的，两面都是在平底锅上烘烤而成的。而也有的蛋糕是在特制的蛋糕盘上烤制的。如果早期的蛋糕是甜味的，那是加一点点蜂蜜调味，或者不加糖烘烤，配着一点蜂蜜食用。在16至17世纪，蛋糕糊中的蜂蜜被糖取代，配料也从蜂蜜变成了糖霜。到了1615年，烹饪书中蛋糕食谱的一大卖点正是“超大的蔗糖用量”。17世纪80年代左右，蛋糕成为常见的饭后甜点，也可以在早上或者下午配着茶或咖啡吃。装饰华丽的蛋糕成为特殊场合或者正式宴会上的特色，例如圣诞节、婚礼和生日。随着糖价的下跌和蔗糖提纯技术的精进，蛋糕中精制糖的分量增加了；糖粉或者糖霜（也叫“糖果制作者的糖”，因为往往是用在糖果制造中）到19世纪已经很普遍了，甚至都有制作蛋糕糖霜的食谱了。

烤蛋糕的传统是由欧洲移民带到美国的，在殖民时期，蛋糕就已经很流行了；时至今日，它还是美国最受欢迎的甜点之一。从简单的姜饼蛋糕、磅蛋糕、天使蛋糕和海绵蛋糕到更丰富的水果蛋糕、芝士蛋糕、糖霜蛋糕和千层蛋糕，甚至是非常繁复的婚礼蛋糕和装饰奇特的杯子蛋糕，这些美国蛋糕都需要加大量的糖。虽然还是有人坚持从面粉开始一点点制作蛋糕，特别是生日蛋糕，但事实上，现在有很多混合好的预拌粉、烤好的蛋糕以及盒装的或者速冻产品出售，供那些不想动手的人去选择。

虽然在 19 世纪初一些黑面包（也叫消化面包）的配方里面有糖蜜，但按照历史传统来看，做面包是不需要加糖的。然而，19 世纪末，面粉厂的石磨被高速滚筒取代之后，一切都大不一样了。小麦中的米麸、胚芽和油脂全被打磨掉了，剩下的只有精制白面。为了给食之无味的白面调味，烘焙师开始加糖，而且随着时间的推移越加越多（糖也可以给面包增加湿度，还能让它保存的时间更长）。19 世纪 80 年代，烹饪书的推荐是八杯面粉混合一茶匙砂糖。十年之后，糖的比例惊人地增加到了一杯面粉一勺糖。商业化的制造者往往会加更多的糖。在 20 世纪，商业面包里糖的用量又进一步增加。相比之下，意大利和法国烘焙师制作面包时只加入少量糖或者不加糖。

甜甜圈

美国甜甜圈可能起源于荷兰、德国或者英国。荷兰人管这种食物叫“olijkoeken”（油蛋糕）或者“oliebollen”。通常的做法是抓取一点甜面团，在手掌上揉圆，直接下到热油锅里。荷兰版本的油炸面团或者油炸“扭结”在美国很受欢迎，但中间有洞的甜甜圈直到19世纪末才普遍出现。据说，这个洞是一个实用的发明，为了让甜甜圈更容易泡在咖啡里。另一种说法是，这个形状有助于面团受热更均匀。

商品化的甜甜圈是在二战后开始大规模销售的。因为有了平价的机器，甜甜圈零售适合特许经营。比如唐恩都乐、多拿屋、卡卡圈坊和温切尔甜甜圈，都是甜甜圈特许经销商。光是唐恩都乐一家，一天就能售出大概640万个甜甜圈（一年23亿个）。

而另一家甜甜圈连锁店则是由蒂姆·霍顿创立的，他本来是一名平平无奇的加拿大曲棍球运动员，他的第一家店是1964年在安大略省的汉米尔顿开的。店里的经典产品是保证新鲜现煮的咖啡、卡布奇诺、甜甜圈和小的油炸面团，但菜单上很快就加入了新品。公司的扩张非常迅猛，很快就成为加拿大最大的快餐公司。1995年，蒂姆·霍顿在美国开了店。汉堡王2014年决定收购这个连锁品牌，这时候蒂姆·霍顿几乎开了4600多家分店，其中845家是在美国或者加拿大以外的其他地方开的。

80% 以上的甜甜圈生意都是外带的，而且在北美，80% 的甜甜圈都是在中午以前卖出去的。它们在形状、大小和口味上五花八门。有酵母甜甜圈和泡打粉甜甜圈，这两者通常都是油炸的；如果你不想摄入太多脂肪，烤箱烤的甜甜圈是个不错的选择。大多数甜甜圈中间都有个洞，而没洞的油炸面团（或者其他仿照这种形状的食物）经常是单独售卖的。有馅甜甜圈的注芯部分一般是果酱（果冻）、卡仕达酱或者诸如此类的其他甜馅；其他的装饰还包括厚厚一层糖霜或者肉桂糖、薄薄的巧克力脆皮、香草、巧克力或者其他口味的淋面、椰子碎或者彩糖。类似的点心还有油炸麻花，可以理解为长条的甜甜圈面团扭在一起之后油炸；还有俾斯麦甜甜圈，是闪电泡芙形状的大号果酱甜甜圈。

恩特曼家的混合口味甜甜圈

冰激凌

冰冻甜食、冰激凌和雪葩——这些冰冻甜点的甜味最开始是因为加了果汁的缘故——可能起源于16世纪的法国或意大利。从17世纪开始，它们在欧洲的咖啡厅里出售，到19世纪，欧洲城市里已经有很多卖这种冰冻甜点的小商贩了。一些冰激凌配方是在18世纪出现在英国烹饪书中的。欧洲移民又把做冰激凌的手艺带到了美洲，18世纪90年代，殖民地的一些城市里就有冰激凌商店开张了。很多冰激凌的配方——大多数添加了大量的糖——是19世纪出现在美国烹饪书中的。

三个主要的冰激凌口味——巧克力、香草、草莓——在19世纪出现，此后一直热度不减。当然，在19世纪里，其他的口味大量涌现，例如用创意十足的配料、果酱和装饰制作的精致冰激凌，以及用冰激凌制作的汽水饮料。20世纪末见证了混合甜品的兴起——高品质的冰激凌嵌满了曲奇块、糖果、巧克力、坚果或者水果，也有的是一圈圈的浓厚焦糖、浓巧克力或者花生黄油酱从冰激凌上流下来。

对于19世纪的人来说，出去享用冰激凌是一个颇为风雅的活动。冰激凌往往在一个叫作“冰室”的特定场所里供应，冰激凌会被舀进精致的小玻璃碗里，用勺子吃。它算是个火爆的夏日单品，原料也很便宜，但街头商贩们主要的烦恼就是如何让产品一直保持低温以及如何不需要碗和勺子也能让人轻松地享用这种美味。而解决之道正是冰激凌甜筒，这是一个 19 世纪末的发明创造。

直到冷藏技术改进，冰激凌商业化生产才得以稳固，使得冰激凌在药店、冷饮柜台和杂货铺里出售成为可能。在美国，冷饮柜台是提供酒精饮料的沙龙和酒吧的主要竞争对手，因此受到了禁酒运动的支持。随着酒吧、沙龙和小酒馆因为禁酒而纷纷关门大吉，冷饮柜台成为人们聚集的场所，冰激凌也声名鹊起。二战结束之后，自助冷柜在杂货店中广泛应用，家里电冰箱的冷冻区也提高了容积和功率，这时袋装的冰激凌成为每日饮食中必不可少的一部分。

到了 20 世纪 50 年代，大的冰激凌制造商通过低价抛售的方式打压小的厂家，超市也变成了全国连锁的品牌。但市场中出现了急需“超高品质”冰激凌的一小批顾客，他们渴望比超市冰激凌含有更高动物脂肪、更少空气的产品。一个运营了几十年的家族产业在 1960 年首次推出了新产品——哈根达斯；由两个来自长岛的年轻人亲手创立的本 & 杰瑞冰激凌于 1978 年在佛蒙特州伯灵顿的一家改装加油站开始贩售。从 1951 年开始，伯雷耶就一直是美国最大的冰激凌厂商。紧随其后的是德雷尔家、艾迪家和蓝铃铛乳业有限公司。虽然冰激凌产业的集中化程度很高，但现在美国最大的一类冰激凌企业其实是私有品牌，通常在当地或者一个地区里贩售。在 2013 年，美国人一共购买了大概价值 110 亿美元的冰激凌——每份都含有大量的糖。

含糖饮料

另一种含糖量很高的食物当属苏打气泡水，跟早餐麦片的发展很类似，它起初也是一种健康食品，结果却截然相反。无论是蒸馏的还是天然带气泡的矿泉水都一直被认为是有一定保健功效的，人为添加了二氧化碳的水也被认为具有药用价值，建在天然泉水上的欧洲温泉度假村里，饮用带气泡的矿泉水是保健养生中很重要的一部分。在 18 世纪的时候，包括约瑟夫·普里斯特利和安托万·洛朗·拉瓦锡在内的几位科学家发现，二氧化碳正是天然泉水、啤酒和香槟产生气泡的原因。普里斯特利还制造出一台可以产生气泡的机器，据说他的发明被献给了约翰·孟塔古，第四代三明治伯爵（正是这个人发明了三明治），他是当时的海军大臣。他还命令普里斯特利在皇家物理学院的师生面前展示自己的发明。普里斯特利服从了命令，而当时的观众里正好有本杰明·富兰克林，他当时就生活在伦敦。

其他的科学家也发明出各自制造苏打水的方法。在 1783 年，约翰·雅各布·施维普改进了制作碳酸水的工艺，在瑞士的日内瓦创办了怡泉公司。在法国大革命期间及其后期，施维普将他的产业转移到了英国，他的苏打水被英国皇室当作药物使用。

到了 1800 年，制造商们发现往水中加入碳酸氢钠溶液可以制出气泡水。而碳酸水则需要在高压下使用硫酸才能生产。操作工人很容易被硫酸烧伤，容器也容易发生爆炸。

1810年后，很多制作碳酸水的机器都被申请了专利，但因为制作工艺很复杂，往往需要培训过的熟练技工进行操作。因为设备的价格很昂贵，做出的饮料被认为有药用价值，所以一般苏打水只是在药店里面贩卖。从碳酸水变成调味苏打水是关键的一小步。人们普遍认为，姜味汽水是在美国第一种商品化出售（1866）的调味碳酸饮料。它的创始人大概是底特律的药剂师詹姆斯·弗诺，他创建的品牌就叫弗诺姜汁汽水。

另外一种早期的软饮是根汁汽水，传统的做法是用树皮、叶子、根茎、香草、香料和其他有芬芳气味的植物调味。在早年间，根汁汽水是一种家里自酿的、有点度数的酒精饮料。后来，这些从植物中获得的萃取物原料被鼓吹为上好的滋补品——那个时期经典的草药。到了19世纪40年代，根汁汽水混合物和糖浆都是在当地制作，然后在糖果铺和杂货铺里贩卖的。冷饮柜台会卖冰激凌和用水果糖浆、糖和苏打水兑的饮料套组，这样的小店在全美遍地开花。

苏打水公司生产出糖浆或者浓缩剂，把这些产品卖给药店，并在那里和碳酸水配套售卖。在1892年，威廉姆·佩因特发明出了锯齿状瓶盖，让瓶装饮料的密封变得简单、便宜还安全，改变了之前的零售模式。在同一时间，瓶子制造的工艺也获得了革新：新型的、更结实的玻璃瓶可以用来装滋滋冒泡的饮料且装瓶过程中不会碎裂。

在禁酒令实施期间，制作和贩卖酒精饮料是违法的，软饮便迎来了新的高峰。在20世纪20年代，连锁快餐店也出现了，几乎所有的快餐店都出售冷饮。当1933年禁酒令被废除的时候，软饮和快餐店已经在美国形成了成熟的体系，并且销售额保持上升趋势。

软饮制造商会在推广活动和广告上花掉几十亿美元。其市场活动的目的是通过动画片、电影、视频、赞助慈善活动和游乐园来吸引小朋友。此外，软饮厂商还很喜欢赞助赛事活动、抽奖和游戏，无论是广播、纸媒还是在网络上，其中大部分针对年轻人。在《液体糖》2005年的研究中，公共利益科学中心（CSPI）揭露称，软饮公司甚至将学校定为目标，投放广告来销量产品，他们还说“人们饮食中超过三分之一的精制糖来自软饮”。CSPI指出，最大的单一精制糖来源就是软饮，其为小男孩和小女孩分别提供了9%和8%的热量。他们的研究报告还显示，美国有75%的青少年每天都喝碳酸饮料。

在海外，美国的碳酸饮料公司也迅速扩张。可口可乐和百事可乐的销量超过了世界碳酸饮料总份额的70%。在全世界，碳酸饮料公司一天差不多可以卖出13亿瓶产品，每瓶正常糖分的饮料大约含有8勺的糖。

能量饮料和运动饮料

在很多其他的饮料中也都会加糖，比如果汁、水果冰、咖啡饮料和能量饮料。从20世纪40年代果汁变成批量生产的大众商品之后，糖就一直被添加到果汁饮料中。

生产商们总是喜欢用“水果”二字来吸引那些想要从易拉罐或者瓶子里喝到点营养的潜在顾客，事实上，很多果汁不过就是水果味道的糖水而已。以水果冰为例，这种产品里一般会有16克左右的糖。其他饮料中添加的糖分更多。一瓶20盎司（600毫升）的维生素水里有33克的糖。每份16盎司（475毫升）的星巴克香草咖啡星冰乐里有67克的糖。而一份中杯的Costa热带水果冻饮里有73克的糖——是一个卡卡圈坊甜甜圈含糖量的7倍之多。

能量饮料也很常见，大多数都含有很多甜味剂。一家英国的制药公司率先于1927年发明了第一款能量饮料，名字翻译过来是“葡萄糖苷”。那是一种糖分很高的气泡饮料，主要是帮助生病的小孩子尽快养好身体。一位英国的药剂师沿用了这个配方，并将成品改名为“葡萄适”，还用“葡萄适助力康复”的标语做起了广告。1983年，该公司将这个产品重新定位为一款能量饮料，还打出了新的标语，“葡萄适，补充失去的能量”。

红牛的主要原料有咖啡因、蔗糖、葡萄糖和一些其他的添加剂，是1987年在欧洲上市的。十年之后，它被引入了美国市场，后来成为美国第一款最受欢迎的能量饮料。红牛掀起了一股饮料模仿热潮，例如激振、怪物能量、无所畏惧、摇滚明星、全速前进和无数其他品牌。大公司也都想从这个市场分走一杯羹，比如安海斯－布希的180、可口可乐的KMX、德尔蒙食品的能量炸弹以及百事可乐的狂飙肾上腺素都是类似的能量饮料。这些产品主打高咖啡因这个卖点，一瓶 16 盎司（475 毫升）的饮料里含有将近 500 毫克咖啡因，当然，里面还有各种形式的蔗糖。到了 21 世纪初，光是美国的市面上就能找到超过 300 种能量饮料品牌——大多数都含有高卡路里的甜味剂。

糖也是很多运动饮料的原料之一——这种饮料是通过增加耐力和恢复体力来提高运动员表现的。第一款运动饮料就是佳得乐，是 1965 年佛罗里达大学的罗伯特·凯德和达娜·夏尔斯设计出来的。佳得乐是一种非碳酸饮料，含有水、电解质和很大剂量的碳水化合物（也就是说有很多糖——每 16 盎司的饮料里就含有 28 克糖）。佳得乐开启了运动饮料产业。为了让运动员适应高强度的比赛和训练，运动饮料的确能提高人体的能量（含糖饮料都能起到类似的效果），但很多运动饮料都是被非运动员喝掉的，这就会造成他们超重的问题。

满足点

早在 20 世纪 20 年代初，人们就知道了甜味的饮料卖得很好。生产商们就想知道到底加多少糖可以达到最大的销售量。关于这个问题的研究开始于 20 世纪 70 年代，心理学家安东尼·斯卡法尼和德莱里·斯普林格做的一个关于导致实验室小白鼠肥胖的实验。他们发现，如果小白鼠只是吃普瑞纳狗粮的话是不会暴食和肥胖的，但如果喂它们果脆圈——一种高糖的早餐麦片，小白鼠很快就发福了。两人又用了其他几种常见的超市食物来重复这个实验——比如花生黄油酱、棉花糖、巧克力棒、甜炼乳和巧克力豆曲奇。小白鼠更喜欢甜的食物，而且只要有机会，它们就会一直不停地吃，直到变得肥胖。后续的实验还表明，如果只给肥胖的小白鼠提供不甜的正常食物，它们往往没什么食欲。

几乎是在同一时间，在马萨诸塞联邦纳蒂克的美国军方实验室中工作的研究员霍华德·莫斯科维兹找到了让作战士兵的军粮配给更可口的办法。他的实验证明，随着糖的增加，士兵对于餐食的喜爱程度明显上升。当然，是在某个点之前，如果超过这个点，糖反而使得食物不那么诱人。正是莫斯科维兹发明了“满足点”这个专业术语，形容糖对于人类吸引力的峰值（满足点也用来描述油脂和盐分的摄入量）。这一系列研究的结论就是，对糖的喜爱是与生俱来的本性，人类对于甜食的喜爱是内在固有的特性。1981年的时候，莫斯科维兹离开了军方实验室，在纽约州的白原市开了自己的咨询公司，很多食品公司的总部都设在这里。他的公司就是帮助这些食品公司寻找他们产品的“满足点”的。他称得上是功成名就，而找他咨询的公司大多数也都非常成功。

另一个做了很多相关实验的机构是莫奈尔化学感官中心，是费城一家由政府机关和大公司资助的非营利研究中心。研究中心发现，小孩子比大人更喜欢吃甜食。后来，科学家们又在研究中发现嗜甜是儿童生理系统的一部分，这让研究员们得以探究儿童食品和饮料中的糖的满足点。世界各地的其他研究进一步证实了这个结论，例如位于伦敦、企业赞助的感官享受科学研究协会（ARISE），也得出了嗜甜是天生的这一结论。

这些研究帮助食品制造商确定了到底需要在产品中加多少糖才能刺激消费。全世界的糖果和麦片制造商、烘焙师和软饮厂商都将自家产品中糖的用量提高到科学界定的满足点水平。甜食和饮料的消费量一路突飞猛进——当然全世界顾客的腰围也随之增长，这自然引起了社会对于食品制造业和快餐业的大量抨击。

[第七章]

糖的阴影

在过去的四个世纪里，人们对于食糖的担忧从未停止过。早期主要关注的是糖的摄入与龋齿（蛀牙），在德国人保罗·亨茨纳的著作中最早提出了这一点，他于 1598 年游览英国时见到了 66 岁的伊丽莎白一世。他描述女王的牙齿是黑色的，并说是英国人过度食用糖而产生的负面影响。医学权威也认可这一点：糖会让牙齿变坏。一位学习民法的博士威廉姆·沃恩还提出了其他谴责糖的理由；在他写的《合理健康建议》（1612）中，提出了糖会让牙齿变黑并腐烂。詹姆斯·哈特在《克林克》（又称《节食之疾》，1633）中宣称不加节制地在糖果、甜点和糖李子中添加糖"对身体有害"，会引起便秘、过度进食、肠道阻塞、龋齿（让牙齿看起来发黑）等问题。他还特别提醒了，年轻人要格外注意吃糖的问题。在以后的几个世纪里，很多作家都对吃糖和龋齿的关系大加议论。比如乔纳森·斯威夫特，他就曾在《优雅和巧妙的谈话全集》（1722）的一个对话中写到，"甜食对牙齿不好"。

美国的医学专家也纷纷提出了他们对于精制糖的忧虑。健康倡导者西尔维斯特·格雷厄姆在他的最后一本著作《人类生命科学讲座》（1839）里就提议禁止精制蔗糖的使用，因为这种东西是会让人上瘾的："一个无可争辩的事实就是，人类不可能习惯于有节制地使用任何能让他们上瘾的东西而不破坏整体自然环境。"很多健康领域的改革派也纷纷支持了格雷厄姆的观点。水浴疗法的推崇者罗素·特拉尔在社论以及自己的健康书籍里猛烈地批评糖：

糖被制成了各式各样的糖果、甜食、含片等等，很多还加上了有毒的色素，还用上了药剂师才会用的麻醉剂。有点脑子的药剂师都会指责以上任何一种不当的使用。原糖本就有诸多杂质；精制、干燥后的糖更是导致便秘的元凶。

雷吉纳德·芒特的彩色石版画，20世纪50年代由英国卫生部发布。

然而，也并不是所有的改革派都同意特拉尔极端的观点。复临安息会的信徒约翰·哈维·凯洛格在密歇根州的巴特克里市创办了疗养院，他就认为童年时期的肠道问题主要是由肉类和糖引起的；他还说美国人对于糖和甜食的偏爱需要被严格控制，因为糖妨碍了人体的正常消化。但他没有支持全面禁糖，只是说人们应该少吃点糖，用蜂蜜、枣和葡萄干作为代替。

毋庸置疑，牙医是糖的反对者，同时，很多医疗机构也都觉得糖是个大麻烦。在 1942 年，美国医学会食品与营养理事会就提出，“出于对公共健康的考虑，应该不惜一切代价限制糖的摄入，特别是那些没有其他高营养成分添加的食物。”医学界也担心糖的摄入，尤其是它对低血糖（血液中糖含量低）的影响。医学博士 E. M. 亚伯拉罕森和 A. W. 泊泽在《身体、大脑和糖》（1951）一书中写道，糖会导致“一系列疾病”，只要不吃糖后，病人的健康状况很快就能得到改善。虽然这本书主要是基于个人经验——亚伯拉罕森是一位治疗糖尿病的医师，他让泊泽服用了“高胰岛素”，但这本书还是名声大噪，卖了 20 多万本，还获得了其他医学专家的认可。一位英国皇家海军的外科医生托马斯·L. 克利夫和南非医生乔治·D. 坎贝尔共同研究了很多个社会中的样本发现，患糖尿病、心脏病、肥胖症、消化性溃疡等慢性病的概率会随着精制糖、白面和白米的摄入量的增加而增加。吃精致碳水化合物的量越少，患上这些疾病的概率也就越低。他们在《糖尿病、冠状动脉血栓形成与糖精病》(1966)一书中发表了这个结论。虽然有的人对他们的观点嗤之以鼻,但大多数医学专家还是会建议他们的病人减少糖的摄入量。

糖的替代品

随着人们对糖尿病和肥胖越来越担心,人们开始发明零卡和低卡的甜味剂。第一种人造的甜味剂是糖精,是一种白色结晶粉末,它比蔗糖的甜度高300~500倍,却不含热量。它是在1879年被坐落于巴尔的摩的约翰·霍普金斯大学的一名硕士生发现的。一些像孟山都这样的公司开始把糖精当成商品贩售,但直到一战实施了糖配给之后,这个产品才火爆起来。

战后,糖精成了糖尿病患者的福音,还被用在减肥人士的节食产品当中。1977年,一份加拿大的研究报告显示糖精在动物实验上有致癌的风险,美国食品药品监督管理局(FDA)宣布在更多研究发表之前暂时禁止这种产品的使用。后续的研究并没有证实先前的结论,所以限令也就在1991年撤销了。

第二种人造的甜味剂是环己基氨基磺酸钙,从1952年开始用于制造低糖苏打水。这种物质的变体也被广泛地运用在不同食品的制造中。20世纪60年代的实验研究显示,环己基氨基磺酸钙可能致癌,在1970年就被美国食品药品监督管理局禁止使用了。到了70年代末,大多数低卡食品

一战时期，号召美国人减少蔗糖摄入的广告。

都是用第三种糖替代品制作的，那就是阿斯巴甜代糖（常常用“NutraSweet”或者“Equal”的品牌名营销）。

甜菊糖是一种从向日葵属的植物中提取出来的天然零卡代糖。这种植物提取物比餐桌上用的糖甜上 300 倍。甜菊糖从 20 世纪 80 年代开始就在日本非常受欢迎，后来也被亚洲和南美的许多国家广泛使用。1994 年，美国食品药品监督管理局将甜菊糖列为草本添加剂，需要在食品标签中标识出来。2008 年，美国食品药品监督管理局又批准了从甜菊糖衍生出来的两种甜味剂：分别是嘉吉公司和可口可乐公司共同研发的“Truvia”，以及百事公司和全球甜味剂公司研制的“PureVia”。几年后，美国食品药品监督管理局将纯净态的甜菊糖列在了“公认安全”的食品行列。

另一种最近的发明是三氯蔗糖,比普通蔗糖甜上600倍。它是在1991年率先由加拿大批准使用，7年之后美国以几种品牌名来进行营销——“Splenda”“SucraPlus”“Candys”“Cuken”和“Nevella”——它们存在于上千种低糖食品里。另一种零卡甜味剂是乙酰磺胺钾，其甜度是蔗糖甜度的200倍。美国和欧盟都批准了这种物质的使用。NutraSweet公司还制造出了一种叫“Neotame”的甜味剂，比蔗糖甜7000~13000倍。虽然它在2002年就被美国食品药品监督管理局批准了，但并没有在市场上广泛普及。几乎没有什么证据表明这些已被批准的人造甜味剂在短期内有负面影响；关于长期的健康风险的说法，仍然存在争论。

美国超市中卖的代糖

空热量

“空热量”一词是指那些热量全部由碳水化合物或者脂肪提供的几乎不含其他营养素的食物，最早在20世纪50年代被人们使用。在空热量食物榜单顶端的就是像糖果、饼干、蛋糕、派、冰激凌、早餐麦片和碳酸饮料这样的甜食和饮料。

约翰·尤德金在1953年创办了伦敦大学的营养学系，他认定糖摄入和许多慢性病之间有明确的联系。到20世纪50年代末，尤德金一直从事将糖从饮食中剔除，以预防冠心病和助力减肥的活动。在1958年，他出版了一本饮食手册——《瘦身这门学问》(1958)，里面建议采用低碳水饮食方法来减肥。尤德金又陆续发表了一些支撑他观点的文章，在20世纪60年代的英国很有影响力。1972年，他发表了《纯净、雪白、致命：关于糖的问题》，这篇长篇论述引发了英国和美国公众的极大兴趣，然而医学界却普遍反对他的观点，因为他们认为饮食中摄入的脂肪而非糖，才是造成心脏病的主要原因。

美国的医学专家说，美国儿童吃了太多加糖的婴儿食品和早餐麦片，这容易造成多动症以及其他一些贯穿一生的健康问题。美国记者以及长寿倡导者威廉姆·达菲出版了畅销书《糖的阴影》(1975)，书中就写了“人类食用精制糖所造成的多重身心痛苦”。他还将蔗糖和海洛因做类比，称其至少跟尼古丁一样容易上瘾，同样有毒。

尽管有这些警告，但美国的人均食糖量还是不断上升，虽然不是都来自蔗糖。大多数的糖分来源是一直都被错误命名的高果糖玉米糖浆（HFCS）。在20世纪50年代，科学家们找到了从玉米中提取淀粉的方法，又进一步把淀粉转成了葡萄糖，最后又通过加入酶的方式将葡萄糖转化成了果糖。虽然是用玉米做的，但商用的HFCS在化学性质上和蔗糖非常类似。HFCS含有45%的葡萄糖和55%的果糖，而蔗糖含有等量的葡萄糖和果糖。HFCS的好处在于，它比蔗糖的甜度要高。但缺点就是，在刚刚被研制出来的时候，价格比蔗糖贵。20世纪70年代这种情况发生了改变，美国的糖价因为对进口糖的限额和关税等问题上涨，政府给予玉米种植者的补助使得玉米价格下跌。美国制造商才纷纷转向在食品中添加HFCS，特别是饮料厂商。后来的研究显示，人体消化吸收HFCS的方式与蔗糖完全一致。大量研究指出，健康问题主要是和精制糖的总摄入量有关，与HFCS无关。

“垃圾食品”一词是指那些高热量的加工食品，特别是甜食、很咸的零食、快餐以及除了热量以外没什么其他营养物质的甜饮料；这个词最早是在20世纪70年代出现的。在之后的十年中，它被公共利益科学中心（CSPI）的主任迈克尔·雅各布森推广开来，兴起了批判高糖饮食的最早的潮流。据CSPI（以及很多其他机构）所说，问题不仅仅是吃垃圾食品那么简单，还有更有营养的食品被他们挤掉的问题。

精制糖是过量摄入热量的重要原因。据英国权威医学杂志《柳叶刀》在2011年发表的研究显示，自1980年以来，全球的肥胖率几乎增长了一倍，“1980年，有4.8%的男人和7.9%的女人属于肥胖人口。在2008年，世界上9.8%的男人和13.8%的女人属于肥胖人口。”据估计，全球有13亿人超过了标准体重——其中有一半人都属于肥胖人口——几乎世界上每个国家的超重的人口数量都在上涨。超重可能会导致高血压、关节炎、不孕症、心脏病、中风、II型糖尿病、出生缺陷、胆囊疾病、痛风、免疫功能受损、肝病、骨关节炎和一些癌症（包括乳腺癌、前列腺癌、食道癌、结直肠癌、子宫内膜癌和肾癌）。

造成肥胖和超重的原因也很多，但总部在苏黎世的金融公司瑞士信托资助的研究中心对研究结果进行审查后，在2013年指出：“虽然医学研究还不能确切断言糖是导致肥胖、II型糖尿病和代谢综合征的主要原因，但最近的医学研究的重心正围绕在这个结论上。”他们还认为，糖符合“成为潜在上瘾物质的标准”。这个问题在美国最严重，有61%的国民都能被归为超重人口。据瑞士信贷研究所的结论显示，美国在这方面每年花费的成本也高到令人难以置信：30%~40%的医疗支出——大概是1万亿美金——都是用于解决与过量摄入糖相关的问题。

尾声

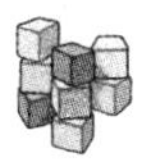

糖生产和消费仍然很容易成为环保、政治和营养健康集团的众矢之的，这也情有可原。环保主义者认定甘蔗种植造成了巴西的雨林破坏、澳大利亚大堡礁的白化以及佛罗里达埃佛湿地的环境恶化。顺着灌溉水流走的化肥和杀虫剂造成甘蔗和甜菜的种植区环境的破坏，以及淡水和海水的污染。在甘蔗种植区的合同工，比如在多米尼加共和国的海地人，受到了糟糕的对待，许多国家已经对甘蔗地里工作的移民劳工问题关注起来，包括美国在内。

公民团体向政府索要了很多农业利益，食品公司也不断游说政府，企图获得本土糖生产的补贴以及国外进口糖的高关税和低进口额度，最近欧盟和美国都通过了这样的政策。这些举措让全世界的糖价都降低，一些世界上最不发达的国家因此产生了严重的经济危机，抬高了发达国家含糖加工食品的成本。

保健专家和营养专家鉴定添加糖是肥胖的一个主要诱因，而且认定它是很多疾病的元凶，例如糖尿病、心脏病、肥胖症、消化性溃疡和其他慢性病。这些人指责大食品公司往自家的产品里放过量的糖，还将目标瞄准小孩，在电视上、收音机里、网上，以及学校和体育赛事现场或附近进行广告宣传活动。

在批评的声浪之中，有的食品厂商开始在产品中减少糖的用量。家乐氏集团和通用磨坊本来生产了市场上 60%

的最不健康谷物片，而且它们在所有谷类食品公司中最关注针对儿童的营销活动，却从 2007 年开始选择降低儿童麦片的含糖量。

虽然反糖运动一直存在，但蔗糖始终是世界上最重要的食物之一：据估计，世界总热量摄入的 8% 都是蔗糖提供的，尽管人跟人之间食用量的差异很大。全球的平均摄入量是一天 17 茶匙（约 70 克）。排在首位的是美国人，他们每天平均要摄入 40 茶匙的糖，每年就是 132 磅（约 60 千克）。紧随其后的是巴西人、阿根廷人、墨西哥人和澳大利亚人，人均每天摄入 30 茶匙。印度人吃的就比较少了，而中国人是世界大国中吃糖最少的，一天只吃 7 茶匙左右，每年也就 4 磅（约 1.8 千克）。

甘蔗和甜菜一直是世界上最重要的农作物。虽然很多国家都会种植它们，但主要产量却相对较少。巴西是当今世界上最大的甘蔗生产国，产量约占全世界作物的 28%，但其中约一半被用来生产乙醇。巴西出口的加工原糖约占全球总量的 25%。生产规模排名第二的是印度，和中国、泰国加起来的糖产量约占世界总产量的三分之一。剩下的则是由其他 114 个国家生产的。

糖会继续在人类的饮食中扮演重要的角色。这不仅仅是因为我们的生理需要，也不是因为垃圾食品和碳酸饮料厂商的成功营销策略在吸引着我们购买甜食和饮料。像糖果、蛋糕、巧克力、冰激凌和碳酸饮料这样的甜食能带给我们愉悦的心情，像小奖励一样帮助我们度过这一天。这些食物总是与美好的时光联系在一起——比如圣诞节、复活节、情人节、万圣节、生日派对和婚礼这样的节日和庆典。在未来，适量食用甜食和饮料仍将是我们生活中不可或缺的一部分。

食谱

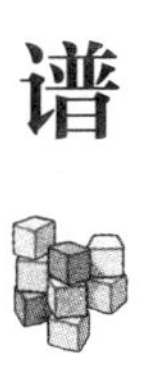

制作杏仁糖

选自《淑女的享受》（伦敦，1611）

取两磅焯熟的杏仁，放进筛子里，在明火上烤干；在石臼中敲击，当杏仁变碎之后，加入两磅打细的糖粉末，加入两三茶匙的玫瑰花水，这可以保持杏仁不出油：当你的杏仁糊变得很细的时候，用擀面杖把它搅得稀一些，让它在一个玻璃器皿的底部沉淀；将容器的一侧抬高一点点，放在烤箱里烤制；再用玫瑰水和糖冷却，然后重新烤制一遍，当你看到冰镇过的物质浮起来变干了之后，就把它们从烤箱里取出来，用精制的装饰物点缀一下，例如用模具刻出来的小鸟和动物图案。在上面笔直插入一些长形糖果，放点饼干之类的就可以端上桌了；你也可以将杏仁糖糊放进模具里，在宴会的时候用。用这种杏仁糖糊，早前的糖匠可以做出信纸、扭结、军队、盾牌、动物、鸟类和其他新奇的玩意儿。

巧克力奶油

选自弗朗索瓦·马西亚洛特的《皇室及贵族料理》

（巴黎，1693）

取一夸脱（约 0.95 升）牛奶和四分之一磅的糖，混合之后煮 15 分钟。将一个打发的蛋黄放进奶油里，煮至有三四个泡沫出现就可以了。离火，和巧克力混合，直到奶油已经有了巧克力的颜色。之后，重新在火上煮沸，过筛，按照喜好享用。

制作美味蛋糕

选自《点心师傅的随行笔记》，又名《厨师、管家、优雅主妇的口袋书》（伦敦，1705）

取12盎司刮得很细腻的甘草、2.5品脱[①]伊索普水、1.5品脱款冬叶水、1.5品脱红玫瑰水、两把迷迭香花朵、一把孔雀草，把这些材料在一口炖锅或者一个可以有盖子的罐子里混合放置三四天，每天摇晃两三次；接着把它们倒进一个平底锅里，小火慢煮两小时，把汤汁挤入一个银色的大盆里，加入一磅粗蔗糖，继续煮至液体变稠，搅成糊状；等你觉得足够黏稠的时候，用小勺舀一点出来，用餐刀搅拌至变凉后，就能知道它的黏稠度是不是足够了；离火之后，应该用一个勺子大力地搅拌到混合物发白，撒上一些焦化的细糖粉，然后把它加入小蛋糕里即可；最好是搅打混合物的时间久一些，不然糖蜜加入小蛋糕后不太美观，很容易就碎了。配方一半的分量就足够制作一次了。

葡式蛋糕

选自爱德华·基德尔的《糕点与烹饪收据：供学者使用》（伦敦，约1720）

取一磅的精制糖、一磅新鲜黄油、五个鸡蛋、一点敲碎的肉豆蔻，在一个平底锅里用你的手将它们搅拌成轻盈、呈现凝固的状态；加入一磅面粉和半磅醋栗，用手非常轻柔

① 品脱是英美制容量单位，英制1品脱和0.5683升，美制1品脱合0.4732升。

地将它们混合；将混合物移到烤盘里，在烤箱里用低温烤制。

你还可以用一样的方法做醋栗籽蛋糕，只需要把醋栗换成醋栗籽就行了。

另一种圣诞曲奇

选自亚梅里亚·西蒙斯的《美式烹饪》第二版

（奥尔巴尼，纽约，1796）

取三磅面粉，撒上一杯研磨得很细的香菜籽，揉进一磅的黄油和 1.5 磅的糖；在一杯牛奶中溶解三茶匙的珍珠灰，倒入面粉里，将混合物揉成 0.75 英寸厚的面饼，捏成或者用模具压出你喜欢的形状和大小，慢慢烤制 15~20 分钟；虽然刚烤出来是干干的，但只要在陶锅、干窖或者阴暗的房间里放上六个月，饼干就会变得更细腻、松软，味道更好。

澄清蔗糖

选自汉娜·格拉斯和玛利亚·威尔森的《完美甜点师》，

又名《管家指南》（伦敦，1800）

将三磅的细糖、糖块或者糖粉放进一个平底锅或者煮锅里；在陶瓷锅里打一个鸡蛋的蛋白，倒入一品脱的清水，用打蛋器将混合物搅打出白色泡沫；将蛋液倒入一个铜质的水壶或者锅里，放在小火上加热；当它开始沸腾的时候，再加一点点水去稀释它，直到浮沫的最上面开始变稠，糖水变得澄清；煮好后，为了让糖液变得更清澈，用湿的餐巾纸或者丝绸筛网过滤一下，滤液放在什么样的容器里都可以，你想用的时候直接舀取即可。

温馨提示：如果糖不是很细腻，在用之前要先煮一下；不然在煮制的过程中浮沫就会因为升得太高而溢出锅。

德式饼干

选自威廉姆·亚历克西斯·嘉林的《意大利甜品师》

（伦敦，1829）

取丁香、肉桂、香菜籽、肉豆蔻各 0.25 盎司，砸碎过纱网备用（或者直接用这四种香料的香精也可以）；取两盎司柠檬皮，一磅甜杏仁磨成粉末（就像做胡桃糖那样）；将这些原料和 24 颗鸡蛋混合，再加入 5 磅糖，加入足量的面粉混合成有延展性的面团。将其揉成正方形、菱形、橄榄形或者其他任何形状；烤制好后，可以根据你的口味撒上一层巧克力糖霜。

关于煮糖

选自M. A. 卡乐美著、约翰·波特主编的《M. A. 卡乐美原创的巴黎皇家甜点指南》（伦敦，1834）

煮澄清的糖会有六种状态：

第一，平滑态。澄清的糖放在火上加热，煮沸几分钟后，取一些在指尖；如果用拇指按压，糖很容易就能分开，而且能拉出一条几乎看不见的细丝，但有点拉扯感，这个迹象就表明你的糖刚好煮到了平滑态。如果是一捏就碎掉了，这就说明糖是在不完全平滑态。

第二，珍珠态。煮制时间比上一个阶段延长一点，还是取一些在指尖，把手指分开将糖拉出一条细线。如果拉出的糖线很容易就碎了，就是不完全珍珠态；如果糖线可以在两个指尖之间拉伸并保持不断，就证明你煮到了完美的珍珠态。

此外，在这个阶段里，煮沸的糖蜜会出现小气泡；气泡会像一颗颗小珍珠一样出现在糖蜜表面。

第三，空气态。继续煮糖蜜，用漏勺从锅里蘸一点糖，在锅上敲一下，向漏勺吹气；如果有细密的气泡通过，就说明糖煮到了空气态。

第四，羽毛态。将糖蜜煮沸，将搬沫器放进锅里，使劲摇晃；糖蜜能很顺利地从搬沫器上被甩下来，形成一张飞舞的糖网，这就是羽毛态。

第五，碎裂态。待糖煮沸一段时间后，将你的手指先泡在冷水里，然后蘸一下糖蜜，再迅速放进冷水，糖应该就能从你的手指上脱落下来。如果用牙一下就能咬碎的话，就是碎裂态；如果是粘在你的牙上，就是煮到了不完全碎裂态。

第六，焦糖态。从第五个状态变成焦糖状态的时间非常短；很快，糖就失去了本来的白色，开始变得有淡淡的颜色，这就证明你的糖变成焦糖了。

糖　果

选自L.-J.布兰切特的《用甜菜制作和提炼糖的工艺手册》

（波士顿，马萨诸塞，1836）

为了毫无遗漏地达成我们强加给自己的目标，不得不谈谈他们是如何制作糖果的；但至少在法国，制作糖果是糖果商的一门手艺，而不是提纯商的，所以，我们只需要笼统地说说工匠制作的步骤即可。

做糖果和面包糖块其实没什么太大区别，只是前者的结晶不是通过搅拌而产生的，而是要靠静静放置。因此，为了让结晶更规则，只能慢慢等待，要避免所有可能导致温度突然下降的原因，在足够长的时间里保持恒定适当的温度。如果不这样做，就会像制作面包糖块一样，产生云雾状的结晶体，这其实是晶体被破坏的样子，就需要刮掉表面重新等待结晶。因此，规则的结晶可以获得糖果；而混乱的结晶过程只能产出糖条。

糖蜜在煮制澄清并且过滤之后，会被重新放进装澄清液的罐子里，再被倒进坩埚里加热至适当的温度。这一步主要取决于你想要的晶体大小，可以根据糖蜜呼吸的强弱判断。

把加热好的糖重新倒进一个半球形的铜盆里，里面要抛光得非常完美。它的直径为15~18英寸，深度为6~8英寸。在盆的边缘下方约两英寸处，每侧都打着八到十个非常小的孔；匠人会拿一根绳子穿过所有的小孔，从一侧到另一侧，一个也不能少；如果怕糖蜜从小孔里面流出来，可以先拿糨糊或者纸，从盆的外面把小孔贴上。

把糖蜜倒到高出线一英寸的位置就算准备好了，立刻把盆转移到一个温室里，温度要高到使结晶过程保持 6~7 天才行。结晶完成之后，把盆从温室里取出来，然后把母液倒掉，也就是盆中仍然是液体的糖蜜。再往盆里加点水，将散落在底部的结晶糖洗干净，把这些水和母液倒在一起即可。

这时在盆底会出现六到九条线厚的结晶底。细线也会被蔗糖晶体包裹，形成花环的样子。把盆在一个方便沥干的瓶子上倒扣；重新拿回温室里，让它回温。两三天之后，糖就彻底干了；从温室中拿出来后，糖应该很容易就能从盆里剥落下来。现在这个状态就可以拿去贩售了。

母液可以拿到工厂去做糖条，比如有杂质的那种或者糖块。

许多糖果呈现出的或多或少的深色调，取决于用来制作它们的糖蜜纯度。如果糖蜜纯度非常高，结晶就是洁白无瑕的。

有的时候他们也会用合适的着色剂以不同方式给糖果上色。如果我们继续探究这些制作细节未免就有点偏离主题了，但这些正是糖果师们全身心投入的制糖艺术。

草莓冰激凌

选自伊丽莎·莱斯利的《烹饪指导》（费城，1837）

取两夸脱成熟的草莓；把草莓蒂摘了，把它们放到一个深盘中，撒上半磅的糖粉。盖上深盘，静置1~2小时。在筛子上将糖渍草莓碾碎，把所有汁水都压出来，再加入半磅的糖粉搅拌均匀；或者按照自己的口味加糖，尽量甜一点，做成糖蜜那样。再加入两夸脱重奶油，用力搅打均匀。把混合物放入冷冻室里，先准备后续的材料。两小时后，把冰激凌放入模具内，或者撒上一点盐和冰之后直接放回冷冻室，进行二次冷冻。再冻两个小时之后就可以吃了。

柠檬汽水

选自《全知道》（伦敦，1856）

四磅糖粉、一盎司橘醋酸或者酒石酸、两德拉克马[1]柠檬提取物，混合均匀。两到三勺这样的混合糖蜜就已经很甜了，倒上气泡水就是一杯可以立即享用的柠檬汽水。

小婴儿食品

选自莎拉·J.哈勒的《哈勒太太新版烹饪书：小家庭实用厨房系统》（费城，1857）

一勺新鲜牛奶同两勺热水混合均匀；用糖块调味，按照自己的口味即可。这个量足以喂一次新生儿了；每两到三小时喂同样的量就可以了，不要太频繁——直到母乳的天然营养跟得上为止。

① 德拉克马是古代药剂师使用的重量单位，1德拉克马约重4.37克。——译注

制作精致糖果

选自安吉丽娜·玛利亚·柯林斯的《最佳西方烹饪全书》

（纽约，1857）

做一些锡制的小模具，认真地刷上油；取一定量的红糖蜜；用漏勺蘸一下糖蜜，然后向孔中吹气，在该州这个步骤叫作“吹气”。如果糖蜜透光就是刚好的状态；往糖蜜里加几滴柠檬香精。如果是制作白色的糖，就等糖蜜稍微冷却之后，在锅中画圆搅拌到有光亮的淋面出现，然后倒进一个漏斗里；将小模具注满糖蜜；等糖放凉、变得坚硬之后，取出放在纸上。如果你想给它们上色，最好是在加热的步骤时完成。

艾弗尔顿太妃软糖

选自M. W. 埃尔斯沃思和F. B. 迪克森的《成功的管家》

（哈里斯堡，宾夕法尼亚，1884）

这是最受欢迎的英式糖果。取三磅最上乘的红糖，在 1.5 品脱的水中煮开，直到糖蜜在冷水中变硬。再加入 1.5 磅甜味新鲜黄油，这可以让糖变软。稍微煮几分钟，等糖重新变硬之后，把它倒在托盘上。如果喜欢的话，可以用柠檬调一下味道。

旋转拉丝糖

选自朱丽叶特·科森的《克森小姐的实用美国菜》

（纽约，1886）

拉丝糖一般用来装饰大块的冰糖水果或者坚果，也可以是牛轧糖；比如上一个菜谱里的太妃糖或者荨麻酒橘子，都可以在脱模之后放上点拉丝糖；或者是用蛋白粘成的马卡龙金字塔或者其他用糖坚果、冰糖水果和马卡龙做成的大型组合装饰品，也可以是在餐桌上立着的糖制工艺品背景。只需将糖蜜煮到“破裂态”，然后像版画中展示的那样，用勺子取一点糖蜜在抹了油的刀上来回来去地拉丝。动作一定要快且稳定；拉丝糖可以做成很长一截，或者短短的样子；也可以直接在被装饰物上进行拉丝。

姜味汽水

选自伊莎贝尔·戈登·科蒂斯的《优秀主妇的家庭料理指南》（芝加哥，伊利诺伊，1909）

取两加仑温水、两磅白糖、两个柠檬、一汤匙酒石乳、一杯酵母、两盎司白姜根在一点水中研磨熬煮出刺激性味道。把混合物倒入一个石头罐子里，在室温中静置 24 小时，然后装瓶。第二天应该就产生气体了。

Sugar

A Global History

Andrew F. Smith

Contents

Prologue

From birth, humans are attracted to sweet-tasting foods, and for good reason: all 10,000 taste buds in the mouth have special receptors for sweetness. Sweet foods cause the taste buds to release neurotransmitters that light up the brain's pleasure centres. The brain responds by producing endocannabinoids, which increase appetite. This may have an evolutionary explanation: about 40 per cent of the calories in breast milk come from lactose, a disaccharide sugar that is readily metabolized into glucose, the body's basic fuel. The sweetness leads infants to eat more, making them more likely to survive.

Naturally bitter plants may signal toxicity, while sweet foods are generally safe to eat and are usually good sources of simple carbohydrates. Once we become conditioned to consume sweet foods, even the sight of them will cause us to salivate; the saliva will help begin the process of breaking down the carbohydrates, signalling to the digestive system that nutrients are on the way.

For millennia, our ancestors cultivated and bred sweet fruits and vegetables and sweetened foods with juice from fruit, berries, figs, dates, nuts and carrots, saps from carob, maple or palm trees, nectar from flowers, and the leaves and seeds of sweet herbs. Over the centuries humans have learned to harvest, refine or concentrate sweeteners such as maltose from grains, glucose from grapes, fructose from fruits, berries and corn, and sucrose from sugar cane and sugar beet. Humans have even harnessed the bee to provide honey, the Old World's first important sweetener.

The most common sweetener for the past 500 years, however, has been table sugar, or sucrose ($C_{12}H_{22}O_{11}$), a disaccharide composed of two monosaccharides—glucose and fructose—that are linked in chemical combination. These separate during digestion; the glucose molecules pass into the bloodstream through the small intestine and are distributed to the organs, where they are metabolized into energy (any surplus not needed for energy is stored in fat cells). Fructose, the sweetest natural sweetener, is mainly metabolized in the liver, where enzymes convert it into glucose.

Most plants contain sucrose, but the greatest concentrations are found in the *Saccharum* genus, a very tall bamboo-like member of the grass family. The genus likely originated in South or Southeast Asia and it consists of several species, each with numerous varieties. Only two species—*Saccharum robustum* and *S. spontaneum*—can propagate in the wild, and they contain comparatively little sugar. *S. robustum* originated on New Guinea, and from it indigenous peoples domesticated *S. officinarum* or Creole cane, which has a higher sugar content than other species. It was such a success that by about 8,000 years ago it had been widely disseminated to the Philippines, Indonesia, India, Southeast Asia and China. In India, *S. officinarum* hybridized with *S. spontaneum*, a cane native to South Asia, to create *S. barberi*, a common sugar cane cultivated in India. In China, *S. officinarum* hybridized again with *S. spontaneum*, this time creating *S. sinense*, a sugar cane commonly grown in southern China.

Humans have cultivated and tapped the sweet juice of various members of the *Saccharum* species for thousands of years, but *S. officinarum* has dominated the sugar cane industry, although other species and varieties have been used for breeding purposes since the late eighteenth century. Growing and processing cane is a labour-intensive activity. All domesticated canes are propagated asexually—sections of the stalk with at least one bud (also called an eye or node) are cut and planted. The cane fields had to be weeded and fertilized, and irrigated in many places. When ripe, the canes had to be cut down. These tasks were accomplished by hand until the invention of mechanical devices in the twentieth century.

Under ideal conditions, cane stalks can grow as much as 5 cm (about 2 inches) per day for several weeks. When mature, they are about 2 inches thick, and they grow to heights of 3.6 to 4.6 metres (about 12 to 15 feet). They reach their optimum sugar content at anywhere from nine to eighteen months. When the stem begins to flower, the sucrose is at its maximum level (ideally 17 per cent). The stalks are cut off just above the root in a process called 'ratooning'. The root then grows a new stalk, which will be lower in sugar content and less resistant to disease; still, stalks can be ratooned a few times before it is more efficient to remove the roots and plant new stem cuttings.

Humankind's dedication to the cultivation of sugar cane clearly demonstrates our millennia-old appreciation of its sweet taste. Initially people consumed the cane juice by simply chewing or sucking on pieces of stalk. It is difficult to preserve or store cut canes for any length of time: once cut, the stalk quickly deteriorates and turns into a brown mush. It is possible to squeeze the juice from the cane, but once exposed to air, it begins to ferment. This characteristic is a definite advantage if the desired end product is alcohol, but not helpful if what is wanted is a sweetener that can be preserved. How our ancestors worked out how to process cane juice so it could be preserved, and how the implementation and improvement of this process affected human history, is the subject of this book.

Chapter 1
Early Sugar History

Extracting the sweet juice from sugar cane and turning it into crystals of sugar is a complicated process. There is little archaeological evidence to indicate just where or when cane juice was first converted into a form that could be preserved for longer periods of time. Most historians consider eastern India, about 2,500 years ago, the point of origin for the sugar industry. The main reason for this attribution is that many early Indian written sources mention cane sugar and its sweet juice. The *Mahābhāshya*, a commentary on Sanskrit grammar attributed to Patanjali and written some time between 400 and 200 BCE, includes recipes for rice pudding, barley meal and fermented beverages—all sweetened with some type of sugar.

Sugar is also mentioned in the *Arthaśātra*, Kautilya's classic Sanskrit work on Machiavellian statecraft dating to 324–300 BCE. This describes different sugar products, from *guda* (the least pure) to *khanda* (the source of the English word 'candy') to *śarkarā*, the purest sugar. The ancient *śarkarā* probably resembled the Indian sweetener, still used today, called jaggery—a coarse, solid sugar that retains some molasses as well as ash and other impurities. (The Sanskrit word *śarkarā* has, ironically, ended up in English as 'saccharin'—a sugar substitute.)

Early sugar products were made by crushing or grinding cane stalks using animal-powered mills fitted with stone wheels similar to those used to grind grain at the time. Crushing expelled the juice, which was then boiled to concentrate it. What's left is raw sugar, which is a sweet but dirty-brown semi-solid that does not ferment. Over time, innovators devised ways to filter out impurities, resulting in a whiter, sweeter and more crystalline product. The crystals could then be removed from the surrounding dark liquid and formed into soft balls. Later they were shaped into solid pieces of hard sugar and eventually these were ground into granulated sugar when needed. The coarse, dark liquid, later called molasses, was removed during the milling process. It could not be further refined into sugar using the technology of the time, but it, too, could be used as a sweetener and for making alcoholic beverages.

The advantages of refined sugar are immense. It can be granulated, pulverized, crystallized, melted, spun, pulled, boiled and moulded. It blends smoothly with other ingredients, either in a home kitchen or on an industrial production line. It can be used to mask the bitterness and enhance the properties of medicines. It is possible to preserve it for a long period of time, making a sweetener available throughout the year. Processed sugar had many culinary uses, such as concealing or enhancing flavours, making alcoholic beverages and preserving fruits and vegetables. Just as important, it could be transported to those regions where sugar cane could not grow, and thus became an important early commercial trade item.

Eastern India, where sugar cane was extensively grown and processed, was also the birthplace of Buddhism. According to Sucheta Mazumdar, author of *Sugar and Society in China* (1998), sugar cane was integrated into Buddhist religious rituals, and many sayings attributed to Buddha (563–483 BCE) include references to sugar cane. Sugary juices were not forbidden to monks observing a fast, and many Buddhist festival foods were made with sugar.

Sugar also appears frequently in other early religious sources, including Hindu works such as the *Buddhaghosa*; or, *Discourses on Moral Consciousness* (*c.* 500 CE), which describes sugar cane mills, the extracted juice, boiling the cane juice, raw sugar and lumps of sugar. Some sugar historians believe that these lumps were pliable, like toffee, rather than hard, while others consider this the first reference to crystalline sugar. Jain literature also mentions a sugar candy, which was particularly important for Jains, who do not consume honey because they believe that it consists of millions of living beings that would die if the honey were eaten.

Presuming that India was the point of origin for sugar manufacturing, it spread quickly to Southeast Asia and southern China. Little is known of the Southeast Asian operations with the exception that sugar—possibly in the form of sculptures—was exported to China by 221 BCE.

There is more information available about the early sugar industry in China, where, legend has it, Buddhist monks introduced sugar cane and the process to make solid sugar. If the monks were not the first to introduce it, they clearly popularized it.

Sucrose, or cane syrup, was not China's first sweetener. In northern China, where grains were the dominant crops, a sweet syrup of maltose was made, mainly from sorghum. Maltose, a disaccharide composed of two glucose molecules, is much less sweet than sucrose. It was the most important sweetener in China and it is still used in Chinese cookery today.

The process for manufacturing sucrose was introduced into southern China by the third century BCE, but it was not commonly used in northern China until centuries later. The Chinese used sugar in medications as well as for sweetening food and beverages; they may have been the first to make rock candy. Cane sugar, however, was not considered a necessity and Chinese sugar processing did not evolve in the way it did in South Asia and, later, the Middle East. According to Marco Polo, who probably visited China at the end of the thirteenth century, the Mongol emperor of China, Kublai Khan, imported Egyptian sugar artisans to help teach the Chinese how to process sugar cane. Indeed, great progress on sugar growing, processing and preparing was made in the Middle East.

Sugar in the Middle East and Mediterranean

The Greeks and Romans visited India in ancient times and became aware of Indian sugar. Nearchos, Alexander the Great's general who, in 327 BCE, sailed from the mouth of the Indus River to the mouth of the Euphrates in Asia Minor, reported in his *Indika* that 'a reed in India brings forth honey without the help of bees, from which an intoxicating drink is made though the plant bears no fruit.'

Small quantities of sugar made their way into the Mediterranean region during Roman times. These imports were used for medicinal purposes. Dioscorides, the first-century CE Greek physician and botanist, wrote in his five-volume *De Materia Medica*, 'There is a kind of concreted honey, called saccharon, found in reeds in India and Arabia Felix', which, he added, has the 'appearance of salt; and, like that, is brittle'. Galen, Seneca, Pliny and others reference a kind of honey imported from India, and many modern observers believe that this was in fact sugar. By the sixth century CE, sugar was shipped from India to a port on the Somali coast and then overland to Alexandria, and from there, small quantities were traded to physicians, who used it for medical purposes.

Sugar cane was grown in Mesopotamia by 600 CE and commercial production began shortly thereafter. The Byzantine historian Theophanes recorded that blocks of *zuchar* were among the booty of great value captured in 622 in the campaigns by Heraklius against the Sassanid

Empire. The Arabs had conquered Mesopotamia by 641, and through them sugar cane and its manufacturing process spread westward to the Nile River, the Nile Delta, the eastern Mediterranean and East Africa. It continued to be disseminated westward to the Mediterranean islands—Cyprus, Malta, Crete, Sicily and Rhodes—and sugar cane was widely grown in northern Africa, reaching southern Morocco by 682 CE. It was later grown in parts of southern Spain, southern Italy and Turkey.

In Mesopotamia the centre of the sugar industry was at the head of the Persian Gulf in the Tigris–Euphrates Delta. Sugar became a very important commodity in Baghdad, which at the time controlled an empire that extended from what is today Iran to Spain. Baghdad, with its estimated population of 1 million, was reportedly the largest city in the world. Ibn Sayyar Al-Warraq's tenth-century Baghdadi cookbook includes scores of recipes that have sugar as an ingredient. The sugar industry thrived until the arrival of the Mongols, who sacked Baghdad in 1258; the region fell into political disorder and sugar production was devastated, but by this time the sugar industry was well established in the Mediterranean.

Growing sugar cane and manufacturing sugar in the Middle East and the Mediterranean entailed heftier investments than in India. The hot, dry climates east of India required large irrigation systems to grow sugar cane. As these systems frequently extended over great distances, governments or very large landowners were needed to construct, maintain and regulate them. Also needed were customers who were willing and able to buy sugar—a very expensive luxury at the time.

Upper Egypt was particularly well positioned to grow sugar cane. With its good warm climate, plenty of water and rich soil in the Nile Delta, sugar became an important ingredient in Egyptian culinary life, at least among the wealthy. On occasion, however, sugar was also distributed to common people. Feasts often included sugar sculptures and guests, depending on their rank, were given between 1 and 25 lb (about 0.45 kg–11.3 kg) of sugar as gifts. Sugar was sold in markets throughout Upper Egypt, which became the major supplier of sugar to the Middle East and Europe. Sugar growers, millers and refiners grew wealthy.

As Europeans re-conquered the Mediterranean lands, such as Crete and Sicily, from the Muslims, they learned how to grow sugar cane and manufacture sugar. During the Crusades Europeans conquered Jerusalem, which they controlled from 1099 to 1187. Sugar production was a lucrative business in this area and Tyre (today in Lebanon) was an important sugar trading city. William of Tyre, who wrote a history of the kingdom of Jerusalem, proclaimed that sugar was a precious product 'very necessary for the use and health of mankind, which is carried by merchants to the most remote countries of the world'. Soldiers and pilgrims in the Near East were introduced to sugar, which they carried back to their home countries. This helped create a demand for sugar in Europe, where monarchs and other nobles, at least, enjoyed it.

Venice, an Italian city-state, had been importing and re-exporting sugar from the eastern Mediterranean since the tenth century. When the Crusades began in 1095, this trade became a very lucrative business. The Venetians expanded their control over Crete and extended their influence over other islands, such as Cyprus. Thanks in part to the sugar re-export business, the small city-state soon became one of the most important powers in the Mediterranean. Although Genoa would later become a central distribution point for Portuguese sugar from the Atlantic islands, it was Venice that dominated the sugar trade in the Mediterranean for almost 500 years.

A serious problem that restricted the growth of sugar production in Europe during the Middle Ages was the lack of labour. This was exacerbated by constant wars in sugar-producing areas of the Mediterranean, followed by the arrival of the Black Death (bubonic plague), which infected Europe from the 1340s. During the next several decades an estimated 30 to 60 per cent of the population of Europe died, creating a labour shortage. In addition, during this time there was an increasing migration from rural areas to cities, which added to the labour shortage in sugar-growing areas. Plantation owners in Sicily and other Mediterranean islands paid premium wages for farm workers, and jobs there were sought after by many Europeans. Still, there were not enough labourers, so plantation owners turned to slaves. Both Christian and Muslims used slaves to plant, harvest and process sugar. At first these were prisoners captured during military campaigns in what is today Bulgaria, Turkey and Greece, but slaves were also acquired from East and, later, West Africa.

Besides a diminished labour pool, Mediterranean sugar manufacturing had another serious limitation—climate. Sugar cane prefers tropical climates. A freeze, or even just a spell of cool weather, could limit the growth of the cane. A more serious problem was the lack of cheap, plentiful fuel to stoke the boilers that converted cane juice into refined sugar. The demand for firewood caused deforestation throughout the sugar-growing areas of the Mediterranean. Deforestation reduced soil fertility and water availability as rainwater flowed away, eroding unprotected soil. The sugar industry began to decline in the eastern Mediterranean—Lebanon, Syria, Egypt and Palestine—beginning in the fifteenth century; by the end of the century these areas were importing sugar. The sugar industry continued to thrive in Cyprus and Crete under Venetian control, and in the western Mediterranean for another century, before it too began to falter.

Yet another change in the eastern Mediterranean sugar trade was the rise of the Ottoman Turks. In 1453, they captured Constantinople, the capital of the Byzantine Empire; they then conquered the Middle East and North Africa and moved into Eastern Europe. The Turks controlled the overland trade routes between the east and the west, and when the trade was disrupted, European royalty and the upper classes were unable to easily import sugar, spices and other riches from Asia. Europeans began to explore ways of circumventing the Turks and Arabs.

Atlantic Sugar

Beginning in the fourteenth century, the Portuguese began exploring the eastern Atlantic, where they found and colonized islands such as Madeira and the nearby island of Porto Santo. Sugar plantations were established on these islands, and sugar was exported from them to Portugal by the midfifteenth century. Any excess not sold in Portugal was exported, generating a considerable profit, which encouraged more exploration and more sugar plantations.

Spain, too, explored the Atlantic and established a colony on the Canary Islands off the coast of northwest Africa. These islands had the advantage of a good climate for growing sugar cane and indigenous peoples who could be enslaved to run the mills. Sugar was exported from the Canaries to Spain by 1500. As was the case in the Mediterranean, lack of fuel was a problem; when the islands were deforested, the sugar industry faltered—and later collapsed due to stiff competition from cheap sugar producers elsewhere.

Optimal locations for growing sugar cane were the uninhabited islands of São Tomé and Príncipe in the Gulf of Guinea, off the coast of tropical Africa; the Portuguese had discovered them in 1470. They had an ideal climate, easy access to slaves in Africa, lots of water to irrigate the cane fields and plenty of fuel to run the mills. Sugar production ramped up, and even with the expenses of the long, arduous sail back to Portugal, it generated large profits for planters.

Chapter 2
New World Sugar to 1900

Christopher Columbus was very familiar with the Atlantic islands and the sugar industry that thrived on them. As an agent for an Italian firm in Genoa, Columbus visited Madeira to purchase sugar in 1478. His first wife's father was the governor of Porto Santo. After Columbus's wife died, he married again, this time to a woman whose family owned a sugar estate on Madeira. When Columbus returned to Spain after his first voyage to the Caribbean, he was convinced that sugar cane would grow on the islands he had explored. On his second voyage to the Caribbean in 1493, Columbus stopped in the Canary Islands and picked up seed cane, which he introduced to the Caribbean island of Hispaniola (today Haiti and the Dominican Republic). Columbus and other Spanish explorers established settlements on other islands, such as Puerto Rico (1508), Jamaica (1509) and Cuba (1511). Sugar cane was planted on these islands, as it would be later in the Spanish and other European colonies of Central and South America.

Hispaniola was the most important New World sugar producer. Sugar was exported from the island to Spain by

1516; within 30 years, the island had 'powerful mills and four horse mills'. Spanish ships picked up 'cargoes of sugar and the skimmings and molasses that are lost would make a great province rich', reported Gonzalo Fernández de Oviedo y Valdés, the contemporary chronicler of the island's history.

While the Caribbean had the perfect climate for growing cane and there was plenty of fuel and water, there was a shortage of manpower. Few Spaniards were willing to migrate to the New World to labour on sugar plantations. Indigenous peoples, such as the Taino and Carib tribes, had no interest in working on these plantations; when the Spanish enslaved them, they were understandably less than industrious. What with constant wars, and epidemics of communicable diseases brought over by the Europeans, an estimated 80 to 90 per cent of the indigenous population of the islands died off during the century following the first European encounter. The Caribbean sugar industry languished.

Brazil was a different story. The Portuguese had landed there in 1500 and later set up small coastal trading posts. It was also an ideal location for growing sugar: the climate was perfect and there was an abundance of fuel for the boilers, plenty of water and an unlimited amount of land. The indigenous people provided a potential supply of slave labour. Small sugar plantations called *engenhos* (the Portuguese word means 'mills', but was applied to the entire sugar plantation complex—fields, mills and factories) were established along the coast by 1520. By 1548, six *engenhos* were operating in Pernambuco; by 1583, there were 66, plus another 36 in nearby Bahia and still others in the southern region.

Portuguese sugar growers are credited with inventing or popularizing several crucial technological improvements. During the early seventeenth century, the *engenhos* adopted a new mill design that crushed cane between three vertically mounted rollers or cylinders. Cane would be fed into two rollers on one side, and then workers on the other side turned the cane back around through other rollers. This was a much more efficient process than the traditional mill press, which was promptly abandoned. The new style of mill could easily be powered by animals, water or even, in some cases, wind; it required fewer workers to operate it and much more sugar was produced as a result.

Yet another important technological change occurred in the process of refining sugar. Traditionally sugar mills had just one large cauldron, in which the cane juice was boiled until supersaturation occurred. The Brazilians created a multiple-cauldron system in which the liquid was ladled from one large cauldron into a series of three successively smaller vessels. This gave overseers much greater control over the process and permitted them to operate on a larger scale.

Brazilian sugar production rapidly escalated, but the industry encountered a major setback when its indigenous labour pool contracted. Disease and wars decimated the native population, and then the Catholic Church in Brazil began to oppose the enslavement of indigenous peoples. A solution soon appeared: the Portuguese sugar colony on São Tomé, which could not compete with the Brazilian sugar industry, shifted its business plan to exporting African slaves to Brazil. Initially many slaves were skilled workers who had worked in the sugar plantations on São Tomé. Later the slaves were any people who could be acquired in Africa and São Tomé served simply as a holding area and a point of departure for Portuguese ships that crisscrossed the Atlantic, transporting slaves to Brazil and elsewhere in the New World, and carrying sugar home to Europe. During the seventeenth century alone, an estimated 560,000 African slaves were shipped to Brazil and other European colonies in the Americas.

Brazilian sugar production intensified and large amounts were exported to Europe. By the late sixteenth century, sugar was Brazil's most important export, exceeding the production of the whole of the rest of the Atlantic world combined. Outpaced by Brazil's output, the Mediterranean sugar industry disappeared altogether and it rapidly declined on the Atlantic islands. Brazil dominated world sugar production.

Sugar Refining

Europeans separated the tasks of growing, processing and refining sugar. Growing and basic processing occurred in their colonies in the Atlantic and Americas, but the refining was accomplished in European cities. This division between producing and refining sugar had several advantages. First, it meant that colonial growing areas did not need local factories, which required large investments, to complete the refining process. Instead these refining centres could be centrally located in Europe's large cities, closer to their ultimate market. Second, transportation of sugar from tropical latitudes by ship was slow, and it was difficult to prevent spoilage en route to the home country. Shipping sugar in a less refined form reduced the risk of spoilage and also allowed European refiners to turn out exactly the kind of product their customers wanted. Finally, by completing the process in European cities, refiners generated some profit for the home country, rather than just for the colony. This final point reflects the economic philosophy of the day—mercantilism. It viewed colonies as places to supply raw materials to the home countries, where manufactured goods would be produced and sold back to its colonies.

Europe's premier refining city in the sixteenth century was Antwerp. Initially it controlled the trade in and refining of sugar from Portuguese and Spanish colonies. Thanks to sugar, Antwerp became Europe's richest and second largest city. It remained in this dominant position until 1576, when Spanish armies sacked it. Antwerp's sugar trade collapsed, as did the city's importance. Other European cities, such as London, Bristol, Bordeaux and Amsterdam, jumped into the void. They launched sugar-refining operations, and wealth flowed to them.

Caribbean Sugar

For almost a century, the Brazilian sugar industry dominated the Atlantic world's sugar trade, but it began to lose market share in the mid-seventeenth century when British, French and Dutch colonists established sugar plantations in the Americas. The Dutch established sugar plantations on the northern coast of South America in what would later become Surinam and the island of Curaçao. In 1630, the Dutch occupied Recife in Pernambuco and other Portuguese settlements in Brazil, which they retained for the next 24 years. The Dutch permitted Sephardic Jews to live and practise their religion openly in these settlements. The Dutch and the Jews became intimate with the growing and production of sugar cane. When the Portuguese retook the Dutch-occupied areas in Brazil, both the Jews and the Dutch left. Some settled in Barbados, a British colony.

Barbados was unoccupied when the British began to settle the island in 1627. Early colonists were mainly small farmers who planned to make their fortunes by growing and curing tobacco. Unfortunately Virginia and other colonies produced more tobacco at less cost. It was the Dutch and Jewish refugees from Brazil who introduced sugar cane to Barbados and taught plantation owners how to convert the cane into sugar. Slaves were imported from Africa to grow the cane and operate the rapidly constructed mills. The island quickly focused on sugar production, as did St Kitts, the Leeward Islands and, later, Jamaica, after its conquest by the

British in 1655.

Similarly the French started sugar colonies on the islands of Martinique and Guadeloupe in 1635, and established plantations on the western part of Hispaniola. In 1697, Spain and France signed the Treaty of Ryswyck, dividing the island of Hispaniola into French and Spanish territories. Over the next 100 years the French colony of Saint-Domingue (today Haiti) became the most productive sugar island in the Caribbean.

Large sugar-cane plantations emerged in the British West Indies. Growers paid their expenses by selling molasses and rum to England or to the English colonies in North America. The sale of these by-products made the islands' prodigious sugar production almost pure profit. Some growers made such huge fortunes in sugar that they installed overseers to run their plantations and then sailed home to England, where they purchased large estates. Sugar also brought wealth to many merchants, refiners, shippers, bankers, insurers, investors and distillers in Britain. By 1760, the city of Bristol alone had twenty sugar refineries that annually processed 831,600 lb (about 377,200 kg) of sugar cane. Sugar barons and their allies in England emerged as a powerful political force that influenced Parliament throughout the eighteenth and early nineteenth centuries. Their financial self-interest diverged from that of their counterparts in the British colonies in North America. The economic and political conflict began with molasses.

Molasses and Rum

To sustain the rapidly expanding slave populations in the West Indies, food and other essentials had to be imported, mostly from British colonies in North America. In return molasses, raw sugar and rum were shipped from the West Indies. Molasses, a by-product of the sugar refining process, was a much cheaper sweetener than crystallized white sugar. It could also be used to make alcoholic beverages, and plantation owners and merchants used it in the production of high-quality rum, which they exported to England or sent to Africa in exchange for slaves.

Rum was also made on the French islands in the Caribbean, but French brandy producers objected to the importation of rum into metropolitan France. The French West Indies sugar industry ended up with a massive excess of molasses. Rather than dump it into the sea, the French government permitted its colonies to sell molasses to anyone who would buy it. The obvious market was the British colonies in North America.

Since molasses from the French West Indies cost 60 to 70 per cent less than the product from the British West Indies, New England colonial ships acquired it in bulk from Martinique, Guadeloupe and Saint-Domingue. New England was ideally suited for rum production: it had skilled workers needed to make the stills, an abundance of ships to transport the bulky molasses and plenty of wood for fuelling stills and making barrels. Rum quickly became the distilled beverage of choice in North America. By 1700, New England was importing more molasses from French colonies than from British ones. In exchange for the molasses and raw sugar, American merchants sent lumber, fish (mainly salt cod for slaves) and other provisions.

As a result of this trade, British West Indian sugar growers lost business, so much so that from 1716 they began to urge the British Parliament to restrict New England's imports of sugar and molasses from French and other European colonies in the Caribbean. Their proposed laws would give the British West Indies a complete monopoly on the molasses and sugar trade, allowing sugar-cane growers to set their own prices and make substantially greater profits. Parliament finally passed the Molasses Act in 1733. The law placed a duty of sixpence per pound on sugar, molasses, rum and spirits imported from non-British colonies.

Had the Molasses Act been enforced, it would likely have crippled New England's fishing and lumber businesses, because these products were traded to Spanish, French and Dutch West Indian possessions. Enforcement would also have hindered the slave trade, in which some New Englanders were involved. But passing the Molasses Act and enforcing it were two different things. The only enforcement provision in the law required the tax to be collected by local customs officials, who were often friends of those engaged in the molasses trade. Customs officials were also few in number and could easily be bribed to look the other way as smugglers shipped in molasses through the thousands of coves along the New England coastline where goods could be landed undetected.

In the 1730s, Parliament permitted growers in the British West Indies to trade sugar directly with countries in Europe. The growers' fortunes improved and they stopped pressing for enforcement of the Molasses Act. Even so, the act remained on the statute books for the next 30 years, during which time molasses was openly smuggled into North America. It was a major misstep to pass the Molasses Act and not enforce it—a blunder that would have serious repercussions later.

American Sugar

In 1725, New York launched its first sugar refinery, and sugar producers in the Caribbean began to ship raw sugar to the city. This partially refined cane sugar was sweeter and more expensive than molasses, but unlike molasses it could be refined into pure sugar. Other refineries were soon established in New York, where they were the largest buildings in the city, and sugar refining became one of the city's most lucrative industries.

As a result of the Seven Years War (1756–63), England acquired vast new possessions in North America—and the ongoing expense of defending them. To generate revenue, Parliament passed the Sugar Act, which lowered the levy on imported molasses to its American colonies that had been imposed 30 years before. As this was a decrease on the existing duty, no one in the British Parliament thought there would be any concern in its American colonies. But the act also included strong enforcement provisions, such as stationing British warships to patrol the coast and sending British custom officials to American ports to enforce the collection of duties.

By the time the Sugar Act was passed, New Englanders had been smuggling molasses and other contraband for 30 years. Enforcement of the Sugar Act made smuggling much more risky. Americans protested the Sugar Act and, in support, many merchants in Boston and New York agreed to stop buying British imports. The Act was repealed, but the colonial resistance set Parliament on a course for more laws to enforce its taxing authority in British North America, and these in turn created even bigger colonial protests. The result was a military conflict, which broke out in 1775.

The American sugar trade and refining businesses collapsed during the war, since the British Navy sporadically controlled the oceans. New York City, the centre of American sugar refining, was occupied by British forces for eight years, and sugar production collapsed. When the war ended, the sugar trade resumed. New York's sugar-refining industry quickly revived and then expanded as raw sugar flooded in from the Caribbean. In 1803, the United States purchased from France the vast tract known as Louisiana. Sugar cane had been grown in the southernmost tip of the Territory—around the Mississippi Delta—since the 1750s, but it was at the northern fringe of the crop's growing range; rainfall was irregular and the growing season was relatively short for sugar cane—only ten months. The cane had to be harvested in the autumn lest freezing temperatures destroy the crop. In 1795, Jean-Etienne Boré, a Frenchman, imported Haitian slaves who were knowledgeable about growing and processing sugar cane. With their help, Boré's plantation did well, and other sugar refineries were constructed. By 1812, the territory had 75 sugar mills in operation. The industry received a major boost in 1817, when ribbon cane was introduced; this plant matured faster than the variety that had previously been planted. The 1820s saw a tremendous growth in sugar-cane cultivation, exploiting the ample supply of slave labour and supported by federal tariffs on imported sugar. Thanks to cheap sugar, America's annual consumption per capita rose from 13 lb (about 5.9 kg) in 1831 to 30 lb (about 13.6 kg) by mid-century.

Throughout the nineteenth century, most American sugar was refined in New York, which was an ideal location for the industry. Its port facilities were the best on the East Coast, facilitating the shipping of raw sugar from the Caribbean and Louisiana. New York itself had a large market for sugar, and the city's road, canal and later train connections meant that refined products could easily be shipped north, south and west. The German-born William Havemeyer, who had been an apprentice at a London sugar refinery, immigrated to America and in 1799 began running the Edmund Seaman & Co. sugar refinery in New York City;

six years later, he opened his own refinery with his brother. It was only one of several refineries operating in the city at the time—and more were soon to be launched. In 1864, the Havemeyer family built the largest and most technologically advanced sugar refinery in the world in Williamsburg on Long Island.

As manufacturing methods improved and production rose dramatically, sugar prices dropped steeply. In 1887, the Havemeyers and seven other sugar industry leaders formed the Sugar Trust; their intention was to curtail production in order to raise prices and profits for all the companies. Following the acquisition of more companies, the resulting conglomerate was named the American Sugar Refining Company. Inefficient plants were closed while others were combined, and American Sugar Refining unofficially but effectively fixed the price of refined sugar. In 1900, the company created a subsidiary, Domino Sugar, to market the parent company's refined sugar. By 1907, the American Sugar Refining Company controlled 97 per cent of all production of refined sugar in America.

Sugar and Slavery

Until the mid-nineteenth century, the entire sugar industry in the Americas was based on slavery. Slaves were acquired from Africa and transported to the Americas to be exchanged for sugar, which was exported to England, where the ships then took on goods to be exchanged for slaves in Africa. The slaves were then transported to the Caribbean to work on sugar plantations. This became known as the 'Triangle Trade'.

Slaves on sugar plantations were subjected to long hours of strenuous work in cane fields, mills and factories. Their lives were cut short by heavy workloads, poor diet, rampant diseases such as yellow fever and the lack of medical care. Some slaves, particularly in Brazil and Jamaica, escaped and formed their own communities in the interior. Workers who died or disappeared had to be replaced, and in the first 75 years of the eighteenth century, the West Indies alone absorbed 1.2 million African slaves. An estimated 252,000 slaves arrived in Barbados and another 662,400 in Jamaica during the period from 1700 to 1810. Slaves soon outnumbered Europeans, particularly on the Caribbean islands. In the French colony of Saint-Domingue, in 1789, a white minority numbering only 32,000 controlled an estimated 500,000 slaves.

Slave revolts, a regular occurrence in Brazil and the Caribbean, were put down violently, with the rebels usually put to death in a gruesome manner. The only successful slave revolt, in Saint-Domingue, began in 1791. Inspired by the ideals of the French Revolution and the Declaration of the Rights of Man, which proclaimed that all men were free and equal, slaves on Saint-Domingue rebelled against their masters. France sent armies to the island to quell the revolt, but the soldiers were defeated by yellow fever and the guerrilla tactics of their opponents. The rebels finally prevailed in 1803, and on 1 January 1804, Haiti became an independent republic, the second in the Americas. Throughout the rebellion, white plantation owners and overseers who were not killed outright fled the colony, some to Louisiana, others to Cuba. Haiti's sugar industry, previously the most productive in the Caribbean, never recovered.

Opposition to slavery grew in Europe and North America in the late eighteenth century. Quakers and others abstained from consuming sugar made with slave labour, but abstention remained an isolated and individual tactic. When Parliament failed to pass the slave trade abolition bill in 1791, British abolitionists joined together to boycott slave-grown sugar. Since one obvious place where sugar was consumed was the tea table, women took an active role in the abstention movement. It developed a broad base of support, attracting as many as 400,000 supporters. Slavery opponents were not just abolitionists: there were also liberal economists, such as Adam Smith, author of *The Wealth of Nations* (1776), who argued that the costs of slavery far outweighed any financial gain. Others were concerned with the undue political power of West Indian planters, who had skewed Parliamentary bills in their favour to the detriment of the British economy throughout the eighteenth century.

The growth of the abolition movement in England encouraged the importation and consumption of slave-free sugar from India. At the time, little Indian sugar was transported to England, but as the abolition movement picked up steam more orders were placed and by the early nineteenth century sugar from India was generally available in England. Quakers established 'Free Produce Societies', which sold Asian sugar.

In the United States, abolitionists tried to grow sugar beet and abstain from purchasing sugar imported from the Caribbean. American Quakers also supported the maple sugar industry as an alternative, and small amounts of maple sugar were produced beginning in the 1780s. In 1789, Philadelphians agreed to buy a given amount of maple sugar at fixed prices to help the industry get off the ground. Quakers especially urged the use of maple sugar to 'reduce by that much the lashings the Negroes have to endure to grow cane sugar to satisfy our gluttony'. Almanacs urged readers to make maple sugar at home because it was sweeter than cane sugar, which was 'mingled with the groans and tears of slavery'. In the 1830s, articles in the *Episcopal Recorder* and the *Colored American* newspaper urged parents to prevent their children from purchasing sweets in confectionery shops because all purchases of sugar supported 'the whole iniquity' of slavery.

Abstention and other abolition efforts finally brought some success. On 3 March 1807, President Thomas Jefferson signed a bill 'to prohibit the importation of slaves into any port or place within the jurisdiction of the United States'. Three weeks later, the British House of Lords passed an Act for the Abolition of the Slave Trade. Slavery continued in the

British West Indies until 1834, in the French colonies until 1848, and in the United States until 1866. Cubans held slaves until 1886 and Brazilians did so until 1888.

Technological improvements—such as the vacuum pan, the centrifuge and the application of steam power to the sugar refineries in the late nineteenth century—made production less labour intensive, but large numbers of labourers were still needed. Freed slaves were unwilling to work in sugar plantations after emancipation. The demand for a workforce led the sugar industry to engage in hiring contract labourers, particularly from India and China, to work the plantations. Eventually hundreds of thousands of contract workers flooded into sugar-growing areas, and many remained after their contracts were completed.

Cuban Sugar

The immediate beneficiary of the end of the sugar industry in Saint- Domingue was Cuba, a Spanish colony. Sugar cane had been grown on the island since 1523, but the industry did not mature because of Spain's policies, which included laws that forbade Cubans from trading with foreign ships and restrictions on importing slaves.

The Cuban sugar industry did not get off the ground until 1762, when the English controlled Havana for ten months during the Seven Years War. During this period the British introduced tens of thousands of slaves into Cuba.

When the war ended, the British withdrew, but Cuban sugar producers demanded liberalized policies: Spain relaxed its laws regarding the importation of slaves and permitted Cubans to trade with foreign countries. More than 18,000 slaves were brought to Cuba during the 1780s, and more than 125,000 during the 1790s and the first decade of the nineteenth century. The Cuban sugar industry thrived, considerable amounts were exported and finished goods from Europe and the United States flowed into Cuba. Still, in 1790, Cuba produced only 15,000 tons of sugar. But this was about to change.

During the slave revolt in Saint-Domingue, Frenchmen from that island moved to Cuba with their slaves and set up sugar plantations and factories. Sugar production in Cuba was enhanced by an improved transportation system; new roads and, later, railways made it possible to move sugar from the refineries to seaports for export. Simultaneously sugar production on other Caribbean islands decreased, largely because of the emancipation of slaves in the British- and French-controlled islands. Cuba retained slavery and quickly became the most cost-efficient sugar producer in the world. Sugar became Cuba's premier export crop, and the United States became its major trading partner. Small sugar mills were closed, and more efficient central factories began to service several cane growers. Cuban sugar received another boost during the American Civil War (1861–5), when sugar plantations in Louisiana were crippled and there was a jump in sugar's price on the world market. From the 1840s to the

1870s, Cuba supplied 25 to 40 per cent of the world's total sugar.

The Cuban sugar industry stagnated during a rebellion on the island that lasted from 1868 to 1878. During the war, many sugar producers left Cuba and some set up shop in the Dominican Republic. After Cuban slaves were emancipated in 1886, many slaves left plantations and refused to work in the sugar industry. Cuba reached out to contract labourers. During the following decades, it absorbed 1.2 million immigrants from Spain, the United States, China, Haiti and other Caribbean islands.

Yet another serious problem was increased competition from abroad in the form of sugar beet, which was grown and converted into refined sugar in Europe and the United States. But large American corporations increasingly invested in Cuban sugar refining. In 1890, they lobbied the United States Congress to pass the McKinley Tariff Act, which eliminated tariffs on refined sugar imported from Cuba. By 1896, the Sugar Trust alone owned nineteen Cuban sugar refineries.

Exports of Cuban sugar to the United States soared, as did American exports of finished goods to Cuba. Production hit 1.1 million tons in 1894. But then sugar-beet growers and American sugar refiners successfully lobbied Congress to levy a 40 per cent increase in duties on imported Cuban sugar. Spain retaliated by placing a tariff on American goods exported to Cuba. The price of Cuban raw sugar dropped sharply, while prices for imported goods from the United States soared. Workers on sugar plantations were laid off, and many joined Cuban guerrilla groups who were fighting for independence from Spain. Guerrillas destroyed sugar refineries as well as cane fields, and the Spanish colonial authorities retaliated with harsh measures, including creating concentration camps, to put down the revolt. The resulting atrocities were covered by many American newspapers; inflammatory articles, which came to be known as 'yellow journalism', swayed U.S. public opinion in favour of the guerrillas.

The Spanish American War and Its Aftermath

In February 1898, an American battleship, the USS *Maine*, exploded and sank in Havana harbour. Although the cause was never determined, the Americans blamed the Spanish. Two months after the explosion, the United States declared war on Spain. In five months of conflict, the American military occupied Cuba, Puerto Rico, Guam and the Philippines. The United States also annexed Hawaii, then controlled by American sugar interests.

After the war, sugar output in Puerto Rico, Hawaii and the Philippines increased to some extent, but Cuban production skyrocketed with the advent of new milling methods, such as the use of water mills, enclosed furnaces, steam engines and improved vacuum pans. American investments in Cuban sugar also soared. By 1919, Americans were estimated to control about 40 per cent of the Cuban sugar industry. Cuban sugar production hit 3.5 million tons by 1925.

Looking Backwards

During the four centuries after Columbus's first voyage to the Caribbean in 1492, the sugar industry had greatly changed. It had shifted from the Mediterranean and the Atlantic islands to the Americas. Plantation labour had shifted from a slave-based force to one based on contract workers. Sugar harvesting, milling and processing had emerged from a largely hand-powered system to an industrial one based on machinery and the latest technology that science could devise. Sugar production had shifted from small plantations and mills to one based on multinational corporations. All of these developments contributed to the steep decline in the price of sugar and the rapid rise in its consumption throughout the world.

Chapter 3
Global Sugar

The beet (*Beta vulgaris*) originated around the Mediterranean, and both the roots and the leaves were widely consumed in Europe and the Middle East beginning in Neolithic times. The Greeks and Romans grew it in their gardens. Medical practitioners prescribed it as the remedy for various ailments. Beets survived as a garden plant during the Middle Ages and they were grown all over Europe by the fifteenth century. Sixteenth-century herbals list several varieties of beet, including pale ones with a sweet taste. The French agronomist Olivier de Serres, in his *Théâtre d'agriculture* (1600), was the first observer to report that the root of the beet 'is counted among the choice foods and the juice which it yields on cooking is like a sugar syrup'.

Beets are a sturdy food crop. Unlike sugar cane, they grow in temperate climates; the hardy plants can withstand droughts and floods. Their growing season is relatively short, so another crop can be planted after they have been harvested. The roots (beetroot) can be dried and preserved for later consumption, and they make excellent fodder for cattle and horses. They became a common agricultural crop in Europe during the seventeenth century.

It was a Prussian chemist named Andreas S. Marggraf who discovered another important property of beetroots. In 1747, Marggraf presented a paper to the Academy of Sciences in Berlin reporting that he had derived small amounts of sucrose from them. The variety of beet that Marggraf used in his experiment, however, yielded very little sugar, and the extraction process was neither practical nor economical. Still, it was promising: if this process could be improved, non-tropical countries would be able to supply their own sugar and would not have to import it. The Prussian government funded research into beet sugar production sporadically for the next several decades.

Marggraf repeated his experiments in 1761 and made enough sugar to produce a few loaves, but the process remained impractical from a commercial standpoint. After Marggraf's death, in 1782, his student Franz Carl Achard continued experimenting with beets and found that some varieties yielded more sugar than others. He perfected Marggraf's process and in 1799 gave Frederick William III, the king of Prussia, a few pounds of sugar crystallized from beetroot. Two years later, the king gave Achard financial assistance to construct a factory in Silesia to test his method. Achard learned a great deal about beets and he is credited with being the first person to extract sugar from them on a commercial basis. He determined that white beetroots contained the most sucrose, and they were subsequently used for breeding purposes. Achard claimed that domestically produced beet sugar could be more economical than imported cane sugar, but he was unable to make a success of his factory.

The beet sugar industry revived during the Napoleonic Wars, when France enforced the Continental System, which prevented goods from Great Britain or its colonies from entering European countries that were controlled by or allied with France. This included imported sugar from the British West Indies, which had previously flooded Europe. In 1791, the slave revolt in the French colony of Saint-Domingue greatly reduced the amount of sugar imported from French colonies in the Caribbean. During wartime, the British blockaded continental ports controlled by France, making it difficult for sugar to be imported from any source. France offered a bounty for the production of sugar from beets, and more than 100 sugar-beet factories were established, mainly in northern France, but also elsewhere in continental Europe. Sugar was successfully extracted, but the industry collapsed when peace was restored in 1815 and cheap sugar once again flowed in from the Caribbean.

But sugar beet was not forgotten. Vilmorin, a French seed producer and a pioneer in the selective breeding of plants, began breeding experiments aimed at increasing the sugar content of beets. In 1837, the company introduced the sugar beet (*B. vulgaris var. altissima*), which had a high sucrose content. Simultaneously new extraction techniques were developed that lowered the cost of producing sugar from beets. The beet sugar industry revived in Germany,

France, Belgium, Austro-Hungary, Russia and Scandinavia. Eventually breeders were able to produce beets with roots that were 20 per cent sucrose.

Even with this increase in sugar content and improved methods of extraction, it was still cheaper to import cane sugar than it was to extract beet sugar. Early support for beet sugar production came from Quakers and abolitionists, who opposed buying cane sugar produced with slave labour. This support died down in England and France when slaves in the Caribbean were emancipated in the mid-nineteenth century. Sugar prices should have escalated with the end of slavery in the Caribbean, but expanding sugar beet and cane production worldwide meant that prices declined throughout the nineteenth century. National governments interested in supporting their sugar-beet industries created policies that favoured in-country beet growers by placing high tariffs and quotas on imported sugar. With governmental intervention, beet sugar cultivation in Europe expanded and hundreds of factories were opened during the late nineteenth century. Governmental support continued through the twentieth century, and Europe became a net exporter of sugar by the end of that century.

Sugar beet was also grown in the Americas, and attempts to convert it into sugar began as early as the 1830s. Successful operations, however, did not get off the ground until the 1870s in the United States and the 1880s in Canada. With governmental protection and support, sugar-beet operations rapidly expanded during the early twentieth century.

Technological advancements, such as mechanical harvesting, increased production and efficiency.

African, Asian and Oceanian Sugar

As sugar-beet cultivation grew in temperate areas of the world, sugar-cane growing and manufacturing expanded in tropical Africa, Asia and the Pacific. Mauritius, a British colony in the Indian Ocean that had been acquired during the Napoleonic Wars, had the perfect tropical climate for growing sugar cane, which was introduced in 1829. As the industry grew, however, it faced serious labour shortages. The British plantation owners' solution was to contract with workers, and tens of thousands came from India. By the mid-nineteenth century, the island was producing 9.4 per cent of the world's sugar, and as sugar production declined in the West Indies, Mauritius became one of Britain's major suppliers. Rather than returning to India after their contract was up, most workers remained on Mauritius. When it became independent in 1975, the majority of its population was of Indian descent. Mauritius continues to grow sugar cane, with the European Union its major buyer.

Sugar cane was also grown in Natal, which today is part of South Africa. The cane growing area expanded into adjacent Zululand (today KwaZulu-Natal) after its annexation in 1887. Once again, the growers' main problem was finding a suitable workforce: Africans were not willing to work under the conditions and wages the growers offered. Workers were bought in India, but they, too, found the work disagreeable and shifted to other employment. Many remained in South Africa and opened other businesses instead of returning to India. Indians were replaced by African migrants from other parts of southern Africa, including large numbers of children, from Natal, KwaZulu and Mozambique.

Sugar cane was introduced into the Pacific islands in prehistoric times by Polynesians and Melanesians, who brought it with them as they set out on long voyages. When they reached new islands, they planted the cane. The British first planted sugar cane in Australia in 1788. It was originally brought to Sydney, but the climate there was too cold to grow it. In the 1860s, sugar production was relaunched. In Queensland, convict labour was used at first; New South Wales hired contract workers from Polynesia. In Queensland, sugar-cane growing was launched by small farmers, but this shifted to plantation-style cane growing in the 1880s, using contract labour from Melanesia. Queensland converted large sections of forested land to agricultural uses, with sugar as a leading crop.

By 1900, the sugar industry was one of America's most important businesses: cane was widely grown in Louisiana and Texas and in the newly acquired territories of Puerto Rico, the Philippines and Hawaii; sugar beet was grown in western states such as Utah and California. Sugar refining played an even greater role in many eastern cities. The most important was the Havemeyer operations in New York. As discussed previously, in 1887, Henry O. Havemeyer devised a partnership with other sugar refiners to be called the Sugar Refineries Company, commonly known as the Sugar Trust.

The American firm Ladd & Co. leased land on Kauai to grow and mill sugar in 1835; others followed its example. Few native Hawaiians wanted to work on sugar plantations, so the growers looked abroad for cheap labour and began to import contract workers, initially Chinese men (women were intentionally excluded). By 1860, Hawaii had 29 sugar plantations, and most of the territory's sugar exports went to the United States. When Mark Twain visited the islands in 1866, he was so impressed with the cane growing that he declared Hawaii 'the king of the sugar world far as astonishing productiveness is concerned'. In 1875, Hawaii and the United States signed a Reciprocity Treaty permitting duty-free importation of Hawaiian sugar.

The following year, Claus Spreckels, a German immigrant who operated a beet sugar factory in California, arrived in Hawaii and immediately made arrangements to purchase most of the Hawaiian sugar. He eventually owned or controlled most of the sugar production in California and on the islands until the 1880s.

By 1882, Chinese labourers totalled 49 per cent of the entire sugar industry's workforce in Hawaii, and Hawaii's political leaders became concerned about the large number of foreigners living there. The following year, Hawaii stopped accepting immigrants from China, and most Chinese workers left. In 1887, American sugar interests forced the king of

Hawaii to agree to a constitution giving them most of the power in the kingdom. With the assistance of U.S. Marines, Euro-American business interests overthrew the monarchy in 1893. They then put pressure on the United States Congress to annex the islands, which it finally did in 1898, during the Spanish—American War.

Labour was needed to work the cane fields and operate the refineries. Okinawan, Korean, Puerto Rican, Portuguese and Filipino nationals came to Hawaii to work in the sugar industry, but the largest group was Japanese immigrants, who had been coming to Hawaii as contract workers since 1865. When their contracts expired, many Japanese labourers remained in Hawaii, and their progeny are the largest single ethnic group on the islands today.

The California and Hawaii Sugar Company (C&H Sugar), formed in 1906, dominated sugar production in Hawaii until the 1930s, when sugar plantations were converted to other uses. Today, only one cane grower remains on the islands. C&H Sugar is now part of American Sugar Refining (Domino Sugar), a company owned by Florida Crystals and the Sugar Cane Growers Cooperative of Florida.

Florida is not an ideal place to grow sugar. Its semitropical climate has occasional freezes, which can destroy the cane crop. The southern part of the state would be preferable, but much of it is occupied by the Everglades National Park. Sugar plantations and mills had been established in eastern Florida in the late nineteenth century. Spain ceded Florida to the United States in 1821, and the sugar-cane industry rapidly expanded, but the plantations stagnated as they could not compete with domestic sugar from Louisiana or cheap imports from the Caribbean.

Sugar cane didn't get a second foothold in Florida until the 1930s, when the U.S. Sugar Corporation was launched in Clewiston. It remained a marginally successful company until 1942, when the Second World War drove up the demand for sugar and sugar cane was planted on even more acres, and this was just the beginning. In 1948, the Army Corp of Engineers began to drain the Everglades and establish an irrigation system to protect populated areas of southern Florida from storm damage. During the next 50 years, more than half of the Everglades was drained, and the reclaimed land became the Everglades Agricultural Area. While anything could have been planted there, the local climate—and federal subsidies—encouraged farmers to grow sugar cane.

When Fidel Castro took control of Cuba in 1959, the U.S. drastically reduced purchases of Cuban sugar. In retaliation, Castro nationalized Cuba's sugar operations, many of which were owned by Americans. The U.S. responded by halting all imports of Cuban sugar in 1961. Many wealthy Cuban sugar growers and processors, such as the Fanjul family, fled Cuba and bought land in southern Florida, where they planted sugar cane. The Fanjuls also acquired sugar plantations in the Dominican Republic.

In 1962, the U.S. Sugar Corporation opened the Bryant Sugar House near West Palm Beach, Florida. It was the world's most modern mill. By the early 1980s, U.S. Sugar was the largest sugar producer in the state and Florida was the number one sugar-producing state in America. Part of the reason for this success was the U.S. Department of Agriculture, which since 1979 has given subsidies and 'non-recourse' loans (which do not have to be repaid) to domestic growers, and imposed import quotas on sugar from the Dominican Republic, Brazil and the Philippines.

Subsidies went disproportionately to the largest cane growers and sugar-beet farmers. In 1991, 42 per cent of the subsidies went to 1 per cent of the growers. As a result of limited competition from abroad and domestic subsidies, American consumers have paid from eight to fourteen cents per pound above the world market price for sugar since 1985–and that's on top of the subsidies they paid for through their taxes. Since sugar is added to many processed foods, these subsidies contribute to a hefty increase in the cost of food. The subsidies, quotas and tariffs did encourage sugar production in the United States: by 2000, 60 per cent of the raw sugar consumed in America was produced from domestic sources.

Since 2008, thanks to the North American Free Trade Association, Mexico can more easily export sugar into the United States and Canada without paying duties. This has sparked an increase in Mexican sugar-cane growing and refining. Today, Mexico is the seventh largest sugar producer in the world and in 2013–14 it supplied about 15 per cent of the sugar that Americans consumed.

War and Revolution

During the Spanish–American War in 1898, the U.S. Army occupied Cuba. In 1903, Cuba gained its independence, and the two countries signed the Reciprocity Agreement, which, among other provisions, provided for a 20 per cent reduction in the tariff on Cuban sugar imported into the United States. Even with the tariff, Cuban sugar continued to outsell sugar produced in America. During the following decade, Cuba became the dominant foreign supplier of sugar to the United States. American companies, such as the Czarnikow Rionda Company and the Cuban Trading Company, acquired Cuban sugar operations. In 1936, Alfonso Fanjul Sr, the heir to these companies, married into a major Cuban sugar family, and the combined resources became the biggest sugar operation in Cuba and one of the largest in the world.

As the American sugar beet industry expanded, sugar prices fell. The Depression hit in 1929, and Congress passed the Smoot-Hawley Tariff Act to protect the domestic sugar industry by raising tariffs on imported sugar. The Cuban sugar industry suffered as prices plummeted. In 1934, the U.S. Congress took control of sugar imports and supported domestic production and refining by passing the Sugar Act (Jones-Costigan Act). After the Second World War, however, Cuban sugar imports picked up again, representing between 25 per cent and 51 per cent of all sugar consumed in the United States.

Cuba nationalized its sugar industry in 1961, and the confiscated plantations became state-run operations. Workers were promised permanent employment, and the government pressured them to achieve production targets, but with no financial incentive, productivity declined. In 1968, the crop was such a disaster that Cuba militarized the harvest to make sure production goals were met.

When the United States stopped importing sugar from Cuba in 1960, the Soviet Union and Eastern European countries picked up the slack: during the next 30 years they bought an estimated 87 per cent of Cuban sugar. When the Soviet Union collapsed in 1991, so did much of the Cuban sugar industry: of the country's 156 sugar mills, 71 closed and 60 per cent of cane fields were converted into vegetable farms or cattle ranches. However, with the invention of a process to make ethanol (a type of alcohol that can fuel vehicles) from sugar cane, the Cuban sugar industry has been revitalized in the twenty-first century.

British Sugar

Until the mid-seventeenth century, Britain was content to import raw sugar from the Mediterranean and the Atlantic islands, bringing it to London and other cities for refining. In the mid-1600s, Britain moved from buying sugar from other European nations to acquiring colonies in the West Indies, including Barbados and Jamaica, which became major sugar producers. In the nineteenth century, British businessmen established sugar-growing and refining operations in other diverse places such as Mauritius in the Indian Ocean, Natal in southern Africa, and Queensland in northeastern Australia.

During the mid-nineteenth century, sugar refining in Britain was dominated by two businessmen: Henry Tate and Abraham Lyle. Tate, a successful grocer in Liverpool, became a partner in the John Wright & Co. sugar refinery in that city. Tate took over the company in 1869 and renamed it Henry Tate & Sons. He opened additional refineries in Liverpool and in Silvertown, London, where sugar cubes were manufactured.

Tate's main competitor was the Scottish businessman Abram Lyle, who with partners acquired the Glebe Sugar Refinery in Greenock, Scotland, in 1865, and six years later constructed a refinery in East London. It produced Golden Syrup, a pale but flavourful liquid sweetener that was used for making preserves, as a sweetener in cooking and as a table syrup. The name Golden Syrup, trademarked in 1904, is believed to be the first British brand so registered. In 1921, these two companies consolidated to become Tate & Lyle, which remained one of Britain's largest sugar refiners until 2010, when it sold its refining operations to the American Refining Company.

Sugar beet was grown in Britain in the late nineteenth century, but the industry did not get off the ground until the First World War, when importing cane sugar became difficult. The industry thrived during the 1920s, but suffered during the Depression. In 1936, Britain nationalized the sugar beet industry and combined several companies into what became British Sugar. In 1991, British Sugar became a subsidiary of Associated British Food (ABF). Today, more sugar is refined in the UK from sugar beet than from imported sugar cane.

Chapter 4
Sugar Uses

For much of sugar's history, the only way to taste it was to suck or chew on a cut stalk of sugar cane and enjoy the sweet juice. For at least the last 2,500 years, people in cane-growing areas have used that liquid and products made from it to sweeten foods and make alcoholic beverages. In ancient times on the Indian subcontinent, cane sugar was added to wine made from dates; fruit juice was sweetened with sugar; and sugar water, sometimes flavoured with herbs, was added to other beverages. The Indian epic *Mahābhārata*, which is attributed to Krishna-Dwaipayana Vyasa (*c.* 400 BCE), mentions sweets made with sugar and *krisara*, a liquid food consisting of five ingredients—milk, ground sesame seeds, rice, sugar and spices. The combination would survive, although with shifts in ingredients and consistency, to become punch (from *panch*, the Sanskrit word for five). At first, these foods and beverages were reserved for wealthy households and special celebrations. By the thirteenth century, however, sugar was plentiful enough in India to be widely available, even to the less prosperous in areas where sugar cane was grown.

By the 1200s, sugar was commonplace in southern and eastern China. The writer Wu Zimu describes, in his *Meng liang lu* (The Past Seems a Dream), seven confectionery shops in Hangzhou in eastern China; they sold coloured, flower-shaped candies, sweet rice porridge, spun sugar, flavoured pastes, musk-flavoured sugar and preserved fruits in sugar syrup. Two Chinese cookbooks of the mid-thirteenth century have survived, and both include recipes for cakes, candies and syrups made with sugar-cane products. Sucheta Mazumdar, the author of *Sugar and Society in China* (1998), calculated that almost 17 per cent of the recipes in one book, and 25 per cent in the other, call for sugar. Both offer recipes for fruits and vegetables preserved in sugar. These were very useful formulas: abundant when in season, fruits and vegetables rot quickly after picking. Preserving them in sugar made it possible to enjoy them when fresh produce was unavailable. Sugar also masked the unpleasant flavours of unripe or overripe fruit. Sugared fruits and vegetables were popular throughout southern China, sold by street vendors and in teahouses and taverns.

Rock candy was made by boiling cane syrup to super-saturation point, then pouring it into moulds and drying it in the sun. There was also a paste of sugar and ground pine nuts (or walnuts) that could be pressed into moulds to create edible sculptures of flowers, animals, birds and fruit. The combination of sugar and ground nuts later evolved into the almond paste called marzipan, the emblematic sweet of the Middle East and Mediterranean.

Sugar cane arrived in the Middle East by the seventh century and quickly became a culinary sensation in Persia, Iraq and Egypt. Ibn Sayyar Al-Warraq's tenth-century Baghdadi cookbook includes more than 80 references to sugar in recipes for wine, sugared almonds and walnuts, cookies, crackers, pudding, chewy nougats and hard candies. Many of these recipes have survived in various forms. Al-Warraq's recipe for *natif*, or nougat, may well have been the inspiration for modern Turkish Delight. Al-Warraq's book also included special recipes for the young, the elderly and travellers. For medicinal purposes, the author noted, sugar was soothing to the throat, chest and stomach, and had other important properties.

Small pies filled with sugar and sweetmeats were popular at this time; another very common treat was a thin pancake folded over several times, saturated with melted butter and sweetened with honey or sugar. Another was *kunáfeh*, thin, finely shredded dough moistened with clarified butter, doused with honey or sugar syrup and baked. It was in the Arab world that sugar began appearing in savoury dishes, such as lamb or mutton stewed with peaches, apricots and jujubes (the fruit of the tropical tree *Ziziphus jujuba*, sometimes called Chinese dates). Sugar was also used to sweeten beverages. The Arabs made a drink called *sharbah* by stirring a sugar syrup infused with rose petals, orange flowers, willow flowers or violets into cold water. Other *sharbahs* were made with raisins or fruit conserves. When ice was added to the mixture, it evolved into the frozen dessert that Europeans called sherbet.

Refined sugar achieved its pinnacle of conspicuous consumption in the households of wealthy Egyptians. According to Adam Mez, author of *Die Renaissance des Islams* (1922), the Vizier Qafur's household consumed 1,000 lb (about 450 kg) of sugar daily in 970 CE; during the following century, one Egyptian vizier's banquet featured 20 tons of sugar sculptures in the form of castles and various animals—elephants, lions and deer. Another feast was ornamented with 50,000 sugar figures, each weighing about 4 lb (about 1.8 kg).

European Sugar Usage

Prior to the ninth century, the small amounts of sugar that arrived in Europe were used for medical purposes. At that time, the humoral system, linking a person's health and temperament to different bodily fluids, had dominated European medical theories for more than a millennium and would continue to do so for almost another 1,000 years. In this system, 'sweetness' was a positive quality, and as sugar was the sweetest substance known, it was considered a sort of miracle drug. In addition to its own healing attributes, it could be combined with other medicines to make them more palatable, and sugar supplied calories—and therefore energy—to anyone who consumed it.

Beginning in the ninth century, Venice imported quantities of molasses, sugar and syrups from Egypt and the eastern Mediterranean. Sugar was re-exported from Venice to the rest of Europe. Northern Italian medical writings mentioned sugar in recipes and formulas beginning in the thirteenth century. For instance, *Tacuinum sanitatis* (Maintenance of Health), based on an eleventh-century Arab health manuscript, listed the pros and cons of sugar:

> It purifies the body, is good for the chest, the kidneys, and the bladder. Dangers: it causes thirst and moves bilious humours. Neutralization of the dangers: with sour pomegranates. Effects: Produces blood that is not bad. It is good for all temperaments, at all ages, in every season and region.

Venetian apothecaries specialized in refining raw sugar and became skilful in making syrups, jams, nut-paste confections, candied violets and a 'celestial water of youth', touted as an elixir for long life. Sugar was given as gifts and it was a must at marriage ceremonies, at which brides were given a box of sweets together with a sugar statuette of a baby.

As sugar became more common over the course of the fourteenth century, it appeared more frequently in cookery manuscripts. In the version of *Le Viandier* from 1300 (a later version of which was attributed to Guillaume Tirel—aka 'Taillevent'), sugar appears only in dishes for the sick. In a version of the same manuscript from 1420, sugar is used in most of the recipes. The early fourteenth-century cookery manuscript *Liber de coquina*, probably written in Naples, called for abundant use of honey, but occasionally replaced it with sugar; in such cases the sugar was mixed in with the other ingredients rather than poured over a finished dish, as honey was. Sugar was used in recipes for broad beans flavoured with spices; rice with almond milk; a *torta* sweetened with both sugar and honey; and a dish made with bitter oranges. A Tuscan recipe collection from the end of the fourteenth century emphasized sugar, giving only a marginal role to honey (in fritters and some desserts). Of the 135 recipes in the manuscript, sugar was an ingredient in 24 per cent, according to Alberto Capatti and Massimo Montanari's *Italian Cuisine: A Cultural History* (2003).

By the fifteenth century, sugar was commonplace in wealthy European households; it was used in sauces, pastry, and confections. Maestro Martino's *Libro de arte coquinaria* (The Art of Cooking), written about 1465, used sugar in great quantities. More than 50 recipes include sugar—custard pies, fish and poultry dishes, potages, a dish of broad (fava) beans, sugar-coated seeds, hot and cold beverages, pan-fried cheese, fritters, macaroni and lasagna, tortes, marzipan, blancmange and preserves. Bartolomeo Platina's *De honesta voluptate et valetudine* (Right Pleasure and Good Health), published in 1474, makes even more use of sugar, as does Bartolomeo Scappi's *Opera dell'arte del cucinare* (The Art and Craft of a Master Cook), published in 1570. As one recipe reported, sugar was 'an excellent accompaniment to everything'.

Sugar continued to be a symbol of opulence and a sign

of wealth. When Henry III, king of France and Poland, visited the city-state of Venice in 1574, sugar was a major component in a banquet given in his honour. The napkins, tablecloths, plates, cutlery—everything on the table—were made of sugar. The setting also boasted 1,250 figures designed by the sculptor Jacopo Sansovino, including a queen on horseback between two tigers, with the coat of arms of France on one and Poland on the other, as well as sugar figures of animals, plants, fruit, kings, popes and saints.

By the early seventeenth century, sugar was widely available throughout much of continental Europe, and on the tables of all but the poor. Naturally this robbed sugar of its cachet for the rich. The Florentine Giovanni Del Turco, in his *Epulario e segreti vari* (1602), complained that earlier cookbook writers had relied too heavily on spices and sugar, which did 'not appeal to the taste of many people'. The number of published recipes requiring sugar began to decline, and smaller quantities of it were used overall in sophisticated cookbooks.

English Sugar

In England, the account books of the household of Henry II (1154–1189) indicate that very small quantities of sugar were used. An entry in an account book from the Countess of Leicester's household records that 55 lb (about 25 kg) of sugar were acquired over a seven-month period in 1265, but the cost remained high. The English love affair with sugar—at least among the nobility—began in the fourteenth century. *The Forme of Cury* (c. 1390), written by Richard II's master cooks, includes numerous recipes with sugar as an ingredient: it went into fritters, custards, pies, sauces, stews, forcemeats, meat, fish, poultry, seafood and game recipes as well as alcoholic beverages, including Cypriot wine, German wine and mead. Within these recipes, sugar was combined with a wide variety of other ingredients, including currants, eggs, cheese, raisins, dates, milk, almond milk, figs, pears, rice, bread and virtually all the spices and herbs then available. Several recipes called for 'Cypriot sugar' as well as other types. Thereafter no feast or banquet was thought complete without sugar. The poem 'The Libelle of Englyshe Polycye', written about 1438, bemoans the importation of commodities from Florence and Venice, with one major exception–sugar: 'And yett. there shulde excepte be ony thynge, *It were but sugre*, truste to my seyinge.' The manuscript 'A Noble Boke off cookry' (*c.* 1480) contains scores of recipes with sugar, including those for beverages, such as one in which it is added to claret. In most cases sugar was used in small quantities and its sweetness did not dominate the recipe.

At most banquets, moreover, few sweet courses were served. This changed by the first decade of the sixteenth century, when the price of sugar dropped and it became affordable to households that were merely prosperous, not necessarily noble or royal. One of the poet Thomas

Newbery's ballads mentions a confectioner's shop selling simnels (buns), cracknels (small cakes baked hard so that they crunched when broken), comfits (sweetmeats) and other products made with sugar. The price continued to drop throughout the century, and by the 1590s, sugar was the prime vehicle for demonstrating high status. The Earl of Hertford gave a banquet for Elizabeth I that included a vast display of decorative foods modelled in flat and three-dimensional sugarwork: 'March-paves, grapes, oisters, muscles, cockles, periwinkles, crabs, lobsters, Apples, pears, and plums of all sorts. Preserves, suckats, jellies, leaches, marmelats, pastry, comfits, of all sorts.' There were also:

> Castles, forts, Ordinance, Drummers, Trumpeters, and soldiers of all sorts in sugar-works. Lions, Unicorns, Bears, Horses, Camels, Bolls, Rams, Dogs, Tigers, Elephants, Antelopes, Dromadaries, Asses, and all other beasts in sugar-works. Eagles, Falcons, Cranes, Bustards, Herons, Hawks, Bitterns, Pheasants, Partridges, Quails, Larks, Sparrows, Pigeons, Cocks, owls, and all that fly, in sugar-works. Snakes, Adders, vipers, frogs, toads, and all kinds of worms, in sugar-work. Mermaids, whales, dolphins, congars, sturgeons, pike, carp, bream, and all sorts of fishes, in sugar-work.

By the early seventeenth century, sugar was almost universally praised in Britain. Francis Bacon even proposed a statue for the 'Inventours of Sugars' in a gallery of important inventors in his utopian novel *Nova Atlantis* (1624). James Heart, author of *Klinke; or, the Diet of the Diseases* (1633), proclaimed that 'Sugar hath now succeeded honie, and is become of farre higher esteem, and is far more pleasing to the palate, and therefore everywhere in frequent use, as well in sickness as in health.'

The French-trained Gervase Markham, chef to England's aristocrats, offered dozens of recipes with sugar in *The English Hus-wife* (1615). These included salads, roasted meats, fish, sauces for turkey and other fowl, preserves, puddings, tarts, sweet and savoury pies, jumbles, cakes, pancakes, fritters, marchpane (marzipan), suckets and many more. The recipes in *The Queens Closet Opened* (1655) made prolific use of sugar. It was called for in preserves, cakes, cheesecakes, pancakes, bread, candy flowers, pumpkin and other pies, tarts, puddings, beans, preserves, salad dressings, alcoholic beverages (such as possets and syllabubs), creams and rudimentary forms of hard candy, as well as in medicinal formulas.

Toward the end of the seventeenth century, sugar lost some of its former charm for the British upper classes. Robert May's *The Accomplished Cook* (1685) mentioned sugar in only two recipes—sauces for meat and fish. John Evelyn's *Acetaria: A Discourse of Sallets* (1699) included a few dozen recipes with sugar in them, but he wrote that '*Sugar* is almost wholly banish'd from all, except the more effeminate Palates, as too much palling, and taking from the grateful Acid now in use, tho' otherwise not totally to be reproved.'

As the English upper classes became disenchanted with sugar, other strata of society discovered its allure, and its consumption surged. What fuelled its renewed popularity was beverages.

Drinking Sugar

Throughout the Middle Ages, the most popular European drink was hypocras or hippocras (probably named for the ancient Roman physician Hippocrates), a mulled or spiced wine commonly consumed at the end of a meal as a digestive. Traditionally hypocras was sweetened with honey. A late medieval French hypocras recipe attributed to the physician Arnoldus of Villanova (c. 1310) includes some sugar. In the Middle Ages, the French cookery manuscript *Le Ménagier de Paris* (*c.* 1393) includes a recipe containing 1¼ lb (about 0.57 kg) of sugar. Recipes for hypocras continued to appear during the next three centuries. An English hypocras recipe from 1692 consisted of 2 quarts (1.9 litres) each of Rhine wine, Canary wine and milk, sweetened with 1½ lb (about 0.68 kg) of sugar.

Hypocras disappeared in the eighteenth century, but by that time many other mixed drinks sweetened with sugar were popular. In England and America, these included flips (beer sweetened with sugar, molasses or honey, and frequently strengthened with rum) and possets (spiced, sweetened hot milk combined with ale or beer), which eventually evolved into eggnog. Syllabubs–spiced milk or cream whipped to a froth with sweet wine or cider and sugar–were spirituous drinks for festive occasions. Shrubs, composed of sweetened citrus juice from oranges, lemons and limes mixed with various spirits, were popular drinks, as were hot toddies, made of spiced and sweetened liquor, and cherry bounce, made from cherry juice and rum. There were iced punches for summer and hot punches for winter. Sangaree, a mixture of wine, sugar and spices, evolved into sangria.

European colonies in the New World produced different alcoholic beverages from sugar and its by-products. In Portuguese Brazil the sugar-based liquor was *cachaça*, a potent distilled alcoholic beverage. Portugal's large and politically powerful brandy industry squelched any possibility of competition from this imported product, so most *cachaça* remained in Brazil. But it was likely Dutch and Jewish immigrants from Brazil who introduced the concept of a sugar-cane-based spirit—and its means of manufacture—into the Caribbean. In the French West Indies it was called *rhum*, a word most likely derived from the English in Barbados, where it was variously called *kill-devil, rumme* and *rumbullion*. Eventually these terms were shortened simply to 'rum'.

Molasses, a by-product of sugar manufacturing, was also used to make mildly alcoholic beverages. West Indian slaves just added water to the molasses, which permitted fermentation. The early eighteenth-century historian Robert Beverley recorded that the 'poorer sort' of British colonists in Virginia used molasses to make a type of beer. It was often flavoured with bran, corn, persimmons, potatoes, pumpkins or even Jerusalem artichokes. In New England, molasses imported from the West Indies was used to make rum. New England was ideally suited for rum production: it had access to the metal and skilled workers needed to make the stills, an abundance of ships to transport the bulky molasses from the Caribbean, and plenty of wood for fuelling the stills and making barrels. Rum quickly became America's alcoholic beverage of choice. It was drunk straight, watered down or mixed with other ingredients–often including sugar. The most popular mixed drink was punch, usually composed of rum, citrus juice and sugar, with myriad variations. Milk punches, made with egg yolks, sugar, rum and nutmeg, were popular at parties and balls. Rum was quite popular in England, while in continental Europe the wine industry successfully fought for laws against its importation.

But it was in three non-alcoholic beverages—chocolate, coffee and tea—that sugar became indispensable throughout Europe, particularly England. The chocolate beverage originated in pre-Columbian Mexico, where it was made by mixing ground chocolate with water and flavouring it with vanilla, chilli peppers, achiote seeds and other ingredients. Since there were no sweeteners in the New World, it was an extremely bitter drink. European colonists, after tasting it, added other spices and sweetened it—at first with honey, and later with sugar.

Chocolate and the equipment needed to make it had

been introduced into Spain from Central America in the early sixteenth century, but it was slow to catch on. From Spain, interest in chocolate spread to Italy, and from there to other European countries. At first, it was flavoured with a variety of New World ingredients and sweetened with honey. As the custom of drinking chocolate spread beyond the elite, sugar replaced honey. Among the earliest European recipes for cocoa is this one, dating from 1631, by the Spanish physician Antonio Colmenero de Ledesma, who wrote in the first treatise on chocolate:

> Take a hundred cacao kernels, two heads of Chili or long peppers, a handful of anise or orjevala, and two of mesachusil or vanilla—or, instead, six Alexandria roses, powdered—two drachms of cinnamon, a dozen almonds and as many hazelnuts, a half pound of white sugar, and annotto enough to color it, and you have the king of chocolates.

Over time, Europeans found that they preferred their chocolate without the exotic flavourings, but sugar remained in the mix. Hot chocolate did not become an important beverage in England until the second half of the seventeenth century: chocolate houses were established in London in the 1650s, and several publications extolled the drink's virtues and provided recipes for it. William Coles reported in his *Adam in Eden* (1657) that chocolate 'may be had in diverse places in London, at reasonable rates' and he added yet another benefit–it was, he claimed, an aphrodisiac. According to William Coles, chocolate had a 'wonderful efficacy for the procreation of children'. The first French recipe for hot chocolate appeared in Francois Massialot's *Le Cuisinier royal et bourgeois* (1693).

Coffee drinking originated in eastern Africa and the Arabian Peninsula. Beginning in the ninth century, it spread to the Middle East. Europeans visiting Turkey and Arab countries wrote about this new beverage, many complaining that it was too bitter. Some Egyptians added sugar to 'correct the bitterness', reported German botanist Johann Vesling, who visited Cairo in the 1630s. Turks opened the first coffeehouses in Europe in the mid-seventeenth century. The first coffee-house in Venice was established in 1629, and soon more opened in other major European cities. When coffeehouses first appeared in Europe, they served their coffee black, with sugar as an optional addition, but it soon became an inseparable companion to coffee.

About the same time that coffee and chocolate began to be appreciated in Europe, tea arrived from East Asia. Tea had been consumed in China for thousands of years, and by the Middle Ages tea leaves were exported from there by overland caravan via the Silk Road to the Middle East, and later to Russia. European explorers and travellers tasted tea while in East Asia, but it wasn't until the Dutch began importing it from China (by 1610) that tea became known in Western Europe. It arrived in England by the mid-seventeenth century, but at first was of interest only to the wealthy. By

1658, tea was being served in coffeehouses, and it quickly became popular. Samuel Pepys, a naval administrator who kept a diary recording the smallest details of his everyday life, reported that he drank his first cup of tea in 1660. Within a few years tea was sold in most London coffeehouses along with coffee, chocolate and sherbet, a version of the Middle Eastern beverage made with flavoured sugar syrup.

Sugar was sometimes added to coffee, tea and chocolate in excessive amounts. In 1671, Philippe Dufour, a Parisian coffee seller, published *De l'usage du caphé, du thé, et du chocolate* (The Manner of Making of Coffee, Tea and Chocolate), which described how each of these beverages was consumed in 'Europe, Asia, Africa, and America'. Dufour recommended adding sugar to coffee, but he complained that some Parisians went overboard with this, until their coffee 'was nothing but a syrup of blackened water'.

The first coffeehouse in England was opened in 1652 by a Turkish merchant. A novelty became a trend and then a craze, and by 1675, London alone reportedly had more than 3,000 coffeehouses, frequented by the city's gentry and well-to-do merchants. While sipping their sweetened coffee, patrons discussed business affairs and politics.

Because of the high price of chocolate, coffee, tea and sugar, British coffeehouses remained the province of the well-to-do; lower classes gathered and drank beer in taverns. Then the British East India Company, a subsidized governmental monopoly, began to import tea in bulk: annual imports increased from 250,000 lb (about 113,000 kg) in 1725 to 24 million lb (about 10.9 million kg) in 1800. As volume increased, the price of tea fell below that of chocolate and coffee. Tea drinking soon outpaced the consumption of chocolate and coffee, and as more tea was imported, it became affordable for the middle class. Tea became England's hot beverage of choice.

During the eighteenth century, the sweetener of choice among the less affluent in England was honey, and for good reason: it was six to ten times cheaper than sugar. During the eighteenth century, sugar was imported from the English colonies in the Caribbean in ever-increasing quantities, and its price nosedived as consumption skyrocketed. At the beginning of the eighteenth century, annual per capita sugar consumption in England was 4.4 lb (about 2 kg) per person. In 1784, duties on imported tea were reduced, which was followed by a sharp increase in the use of tea. By the century's end, per capita sugar consumption had increased by almost 600 per cent, to 24 lb (about 10.9 kg). Even the poorest drank tea with sugar, though they could put little else on the table.

An Essential Ingredient in Cookery

Once the price of sugar dropped below that of honey, it began to be used not just as a sweetener for drinks, but as a cooking ingredient. English cookbooks published in the eighteenth century incorporated sugar into many recipes.

Hannah Glasse published her *Complete Confectioner* in 1760—the first such book published in England—and confidently called for sugar in almost every recipe. There were ice creams, ices, creams, conserves, compotes, marmalades, syrups, jams, cakes, icings, breads, biscuits, beverages, candies, wafers, jumbles, timbales, puffs and tarts, as well as directions for making sugar sculptures and preserving fruit, vegetables, berries, spices, nuts, seeds, roots and flowers. Elizabeth Raffald's *Experienced Housekeeper* (1769) contains more than 100 recipes containing sugar: sauces, pastes, pies, fritters, pancakes, gruels, puddings, dumplings, sweetmeats, custards, pastries, spun sugar, floating islands and all sorts of beverages—syllabubs, ales, various wines, possetts, sherbets, shrubs, brandy and lemonade. Sugar was no longer a luxury—it was an essential ingredient.

Just as sugar dominated English cookery, so it dominated American dishes, although refined sugar was expensive in colonial America. Far cheaper was molasses, which could be used both as a sweetener and as the basic ingredient in rum. Molasses was the primary sweetener for cookies, cakes, pies and puddings, but also in mush, vegetable dishes and meat cookery, especially pork. An English traveller in the 1780s complained that Americans served molasses at every meal, 'even eating it with greasy pork'.

Sugar was sold in many forms, the most common being 8- to 10-lb (about 3.6–4.5 kg) cone-shaped 'loaves'. The wealthy purchased large quantities of sugar, but a middle-class family might make one loaf last a year. It has been estimated that as late as 1788, annual U.S. per capita sugar consumption averaged only about 5 lb (about 2.3 kg). Despite this rapid increase in the amount of sugar consumed throughout the world, the love affair with sugar had barely begun. The amount of sugar would increase in beverages, meats, pies and cakes, but particularly in sweets and candies.

Chapter 5
Sweets and Candies

When sufficient sugar is added to certain foods, whether infused by cooking or applied as a coating, it functions as a preservative by inhibiting the activity of microorganisms. This quality made it possible for traders to carry products such as candied orange peel and sugar-coated almonds over long distances. Solid chunks of sugar (both rock candy and loaf sugar) could also easily be traded. It was through such trade that sweets and candies were introduced in areas where sugar cane was not grown, and where particular fruits and other ingredients were not available.

Through trade routes from southern Asia, sugar confections had reached the Middle East by the seventh century, later spreading to Europe. These early confections, such as comfits, pastes, marzipan, pastilles and rock candy, were the point of origin for many present-day sweets and candies, and traces of their centuries-old heritage still linger today. Early European confectionery traditions have evolved into modern sweets—Jordan almonds, marmalade, sweet pies, cake icing, taffy, toffee, bonbons, jawbreakers (gobstoppers), lemon drops, M&M's, Good & Plenty and ice cream—to name just a few.

Comfits (from the French *confit*, meaning 'candied'—*confetti* in Italian) were initially sugar-coated medicines. Doctors and other healers used the miraculously sweet substance to coat the bitter seeds, nuts, roots, spices, herbs and vegetable extracts they prescribed for various ills, doubtless helping the medicine go down. A sick person might need extra calories, which were easily supplied by sugar; depending on the number of comfits consumed, they may also have given a weak patient a little burst of energy.

Comfits that included candied aromatic seeds, such as anise, coriander, cloves, caraway or cinnamon, were common. In India (and in Indian restaurants elsewhere), plain or candy-coated fennel seeds may still be offered at the conclusion of a meal to aid digestion and freshen the breath. Comfits also included sugar-coated nuggets of candy flavoured with an extract from the roots of plants of the genus *Glycyrrhiza*, a small, leguminous shrub native to Europe, Asia and the Americas. The roots have a sweet, anise-like flavour, and a confection was made by squeezing their juice and then cooking it down to thicken it. Called 'liquorice', it became an important sweet throughout Europe from the Middle Ages onwards.

Today, liquorice candy is manufactured in many shapes and flavours throughout the world, although in many cases the actual root extract has been replaced with anise and artificial flavouring. In the United States, the most famous liquoriceflavoured candy is Good & Plenty—little beads of chewy liquorice with a pink or white candy coating reminiscent of traditional Indian comfits. It was first made in 1893. The artificially flavoured liquorice twists called Twizzlers were first sold in 1929. Today, most liquorice sold in the U.S. is mass-produced using synthetic ingredients, but elsewhere, notably in the Netherlands and Scandinavia, real liquorice in various shapes, hard or soft, from mildly sweet to pungently salty, is practically the national snack.

Another common comfit in the Middle Ages was the sugar-coated nut, which also originated in the Middle East and was later carried to Europe. The French term *dragées* describes spiced, sugar-covered nuts, especially almonds. Today, these have been commercialized under a variety of names, such as Jordan almonds, also known as *mlabas* in the Middle East and *koufeta* in Greece.

Candied fruit and citrus peel also arrived in Europe during the Middle Ages and were typically served after a meal at times of the year when fresh fruit was unavailable. They survive in a variety of forms—candied or glazed fruit, chocolate-dipped cherries, jams, marmalades and jellies.

Marzipan (or marchpane in English), a thick paste of ground almonds and sugar, was a popular delicacy in the Middle East by the Middle Ages. It may have originated in Iran, and arrived in Europe through Arab influence. The first located European reference to it is in northern Italy in the late thirteenth century. It's likely, however, that it had been a popular confection, probably made initially with honey, in Spain, Catalonia and Italy well before that. Marzipan was widely adopted in France, Germany, the Netherlands, northern Europe and England. It wasn't just another bite-sized sweet—it was also shaped into figures, such as pigs or eggs, and was often given as gifts on special occasions, such as Christmas, Easter and at weddings. Marzipan remains popular in Europe and in many former European colonies.

Pulled sugar also arrived in Europe in the Middle Ages, likely from Arab sources in the Middle East. It is made by melting sugar with water, then kneading it to form a plastic substance that can be stretched and pulled into various shapes, such as ribbons, flowers or leaves.

During the Middle Ages sugar was also used as a post-prandial *digestif*—in sugared and spiced wine served with fruit at the end of the meal, after the dishes had been removed. This course came to be called dessert (from the French *desservir*, meaning to clear the table). By the eighteenth century, dessert had become an elaborate course that might include creams, jellies, tarts, pies and sweet puddings. Although typically served after dinner, such dishes could also be offered in the afternoon or evening, unconnected with a meal. Dessert-making became a household art, and was usually done by a professional confectioner or by servants under direction of a professional.

Laura Mason, in her book *Sugar-plums and Sherbet* (1998), writes that, by the eighteenth century, English confectioners sold preserved and candied fruit, biscuits, cakes, macaroons, syrups, comfits, pies, tarts and decorative figures made of sugar. Confectioners also imported and sold sweets from other countries. Bonbons (various types of fancy candies that were initially served at the French court) were imported into England and other European countries. These choice treats, consisting of fondant, fruit or nuts typically coated with chocolate, were a luxury for the leisure class–the only people who could afford them–but this was about to change.

Confectioners

By the seventeenth century, comfit-makers and confectioners were selling their sweets at shops for serving at home. Over time, retailers became more sophisticated. In Paris, *limonadiers* sold beverages such as lemonade (hence their name). One *limonadier*, a Sicilian named Francesco Procopio dei Coltelli, opened a café in Paris in 1686. In addition to serving coffee, the café also offered sugar-coated fruit, ices, sugar-sweetened cold beverages, liquors and hot chocolate.

Declining prices for sugar in the eighteenth and nineteenth centuries made sweets more affordable. In England, the number of sweet shops in provincial cities increased fourfold between the 1780s and the 1820s. They sold a variety of candies, some imported from other countries, some sent from London.

Particular types of sweets and candies emerged. Rock candy was made by allowing a supersaturated solution of sugar and water to crystallize over time; hard candies were made by boiling sugar and water to a syrup and then pouring it into moulds or shaping it by hand. These evolved into gobstoppers, lollipops, peppermints and candy canes. Sugared nuts were the starting point for brittles—whole or broken nuts embedded in a sheet of buttery hard candy. Fruit drops, originally made from boiled sugar flavoured with actual fruit juice, are still popular, although today many are made with corn syrup and artificial flavours and colourings that imitate the taste and appearance of fruit. Life Savers are the most famous modern incarnation of the fruit drop in the U.S. and Canada.

Taffy and Toffee

The first references to taffy and toffee appeared in the early nineteenth century in northern England, where the candy was made in homes and confectionery shops. (The candy-makers of Everton, today a district of Liverpool, were famous for their toffee.) The basic recipe calls for sugar or molasses to be boiled with butter and flavourings: orange, lemon, chocolate or vanilla. From this basic recipe, two different sweets emerged.

Taffy is made by removing the mixture from the fire at the 'hard ball' stage—when a drop of syrup instantly forms a rigid ball when dropped into cold water. The candy is cooled a little, then stretched and pulled by hand with the help of a metal hook until it is satiny smooth. In the nineteenth century, 'taffy-pulls' were popular at parties: guests would pair off to stretch the strands with buttered hands, and then enjoy the fruits of their labour.

To make toffee, the mixture is boiled to the 'hard crack' stage, at which a drop of the boiling syrup forms brittle threads when dropped into cold water. When cooled, this produces a dense candy that snaps when broken. In England, toffee was associated with Guy Fawkes Night (5 November, also known as 'Bonfire Night') and was sold under the name 'Bonfire Toffee'.

Taffy and toffee migrated from England to the United States by the 1840s, and both were popularized in East Coast cities, particularly Philadelphia and Atlantic City. John Ross Edmiston, a Pennsylvanian, was probably the first to sell 'saltwater taffy', supposedly created when a storm brought seawater flooding into his Atlantic City candy shop. There was actually no difference in the recipes for regular taffy and saltwater taffy, but the name caught on. Others perfected its formula and expanded the product line by making the candy in a range of appealing pastel colours, various flavours and diverse shapes. By the 1920s, more than 450 companies, many in seaside resorts, were manufacturing saltwater taffy in America.

Manufacturers

Handmade sweets had been sold commercially in

Europe and North America since the late eighteenth century, but they were not widely consumed until the nineteenth century, when the price of sugar declined and the technology of refining it had become more efficient. Sweets were first mass-produced in England in the 1850s, but manufacturing quickly spread to other countries. Sweets appeared in ever-larger quantities and in more varied shapes and sizes as the century progressed. By the late nineteenth century, hundreds of commercial manufacturers of confectionery were operating in the Middle East, Europe and North America. Most produced small, hard sweets, usually retailed in shops, where they were displayed in big glass jars and sold for pennies.

Soft and chewy candies were also mass-produced. One popular confection that derived from the Middle East was Turkish Delight, or *rahat loukoum* ('rest for the throat'), made from sugar (originally honey), a starch or gelling agent such as gum arabic, and flavouring, often rosewater or orange-flower water. Chopped nuts, such as almonds, pistachios or hazelnuts, or pieces of dried fruit, may be added. The cooked mixture is cooled in a pan and then cut into squares and dredged in icing (powdered) sugar. Its invention is attributed to a Turkish confectioner of the mid-eighteenth century. The popularity of this sweet grew throughout the Middle East and Europe, especially Great Britain, where a Turkish Delight chocolate bar, with a rose-flavoured filling enrobed in milk chocolate, has been made since 1914.

Jellybeans—small, bean-shaped sugar confections with a

firm jelly centre and a hard outer coating—may have derived from Turkish Delight. They come in different colours with various associated fruit flavours. The earliest located reference to jellybeans in print appears in an advertisement dated 1886 from Illinois, where they were touted as a Christmas candy. Jellybeans were commonly sold from jars in sweet shops, or in vending machines. It was not until the 1930s that jellybeans were also marketed as an Easter candy, presumably because of their egg-like appearance.

Albert and Gustav Goelitz, German immigrants who opened a candy store in Belleville, Illinois, in 1869, could be called the forefathers of the modern jellybean. By the turn of the century the family business was specializing in buttercream candies, including candy corn, a three-coloured (yellow, white and orange) candy mimicking an enlarged corn kernel. In 1976, Goelitz descendants created the 'gourmet' jellybean, smaller than the standard size and featuring unexpected flavours such as pear, watermelon, root beer and buttered popcorn (reportedly the most popular flavour). They named the new product Jelly Bellies, and now offer 50 flavours, including cappuccino, chilli-mango and piña colada. Today, the company makes Bertie Bott's Every Flavour Beans, named after a product mentioned in J. K. Rowling's Harry Potter books, and Sports Beans, with added vitamin C and electrolytes.

Chewy sweets date back at least to the Middle Ages, where they were made in the Middle East. Among the first commercial chewy candies were jujubes, named for the juju gum (derived from a shrub of the *Ziziphus* species) that was the main ingredient. Today these fruit-flavoured pastilles are made from potato starch, gum and sugar or another sweetener. Another chewy candy, fruit- and vegetable-shaped jujyfruits, soon emerged. The gummy bear was developed in Germany during the 1920s. Animal-derived gelatin is the basic ingredient in these colourful little figures. In 1982, the German candy company Haribo first marketed their 'gummi' candy in the United States. Trolli, another German manufacturer, introduced gummy worms during the 1980s and they have remained popular ever since. Swedish Fish, another gummy favourite imported from Sweden since the 1960s, are made without animal gelatin. Gummy candies are made in hundreds of shapes and flavours throughout the world.

Holiday Sweets

Many sweets and candies are associated with holidays, particularly Christmas, Chanukah, Easter, Halloween and Valentine's Day. In periods when sugar was rare and costly, the less affluent would have eaten sweets only on such special occasions.

The tradition of Christmas fruitcakes dates back to the Middle Ages. Historically, such cakes were sweetened with candied fruit stirred into the batter or dough; by the sixteenth century, sugar was a basic ingredient, and a sugary icing was a common addition. Particular traditions emerged in different areas, such as the British Christmas cake and the German *Stollen*.

Cultures that celebrate the twelve days of Christmas (from 25 December through to 6 January) have their own traditional cakes. Twelfth Night, or Epiphany, is also Three Kings' Day; in France, a *galette des roix* (kings' cake) is served. In Spain and Latin America, the traditional pastry is the ring-shaped *rosca de reyes* (kings' ring), lavishly decorated with candied fruit. Different kinds of 'kings' cakes' are served on Epiphany in many countries; a special king's cake is also a Mardi Gras speciality in some places.

Many different types of sweet, such as butterscotch, chocolate, lemon, creams, caramels, jellybeans and others, were popular at Christmastime in various times and places. It was not until the mid-nineteenth century that the candy cane–a red-and-white striped candy stick with a crook at the top—became part of American Christmas celebrations. The candy cane's invention is attributed to August Imgard of Wooster, Ohio, who purportedly made and decorated Christmas trees with paper ornaments and candy canes. They were not an immediate commercial success, although people made candy canes at home on a small scale. They were tricky to make and, because of their fragility, difficult to ship. This changed in the 1950s, when candy cane machines automated production and packaging innovations made it possible for the candy to reach its destination unbroken. Candy giants including Mars, Hershey and Nestlé now make their own brands of candy cane.

Halloween (or All Hallows' Eve–the night before All Saints' Day) is observed, mainly in English-speaking countries, on the night of 31 October, when children dressed in costumes go from house to house asking for sweets. Since Halloween comes during the apple season in many countries, it has long been celebrated with candied, toffee or caramel apples. Homemade sweets, such as candied popcorn balls and taffy, gradually gave way to commercial candy, notably candy corn (tricoloured kernel-shaped sweets), which was introduced in the 1880s. Today, miniature boxes of popular candies and small, individually wrapped versions of favourite candy bars are specially packaged for Halloween distribution. This Americanized, candy-filled version of Halloween has recently been adopted in other countries.

Fertility symbols employed in pagan celebrations of spring—rabbits, eggs and chicks—were absorbed into Christian celebrations of Easter. Giving Easter eggs to poor children was a tradition that began in medieval Europe. Easter candy, though, is a relatively recent tradition that may have originated in Eastern Europe. The first located reference to chocolate Easter eggs dates to 1820 in Italy. During the 1930s, Easter sweets, such as jellybeans and chocolate bunnies, became part of the Easter basket tradition. The American candy manufacturer Just Born Company began to make three-dimensional marshmallow Easter chicks, called Peeps, in 1953. In 2012, Americans spent more than $2.3 billion on Easter candies, including 90 million chocolate bunnies, 700 million marshmallow Peeps and 16 billion jellybeans.

Chanukah, the eight-day Festival of Lights, celebrates a Jewish military victory over the Seleucid Empire in 164 BCE and the rededication of the Second Temple in Jerusalem. Chanukah is celebrated at home, with modest presents for children, one for each night. Small coins, or *gelt*, were the usual gift. In the 1920s, urban candymakers began promoting their wares as ideal Chanukah gifts. New York's Loft Candy Company, for instance, sold round, flat chocolates wrapped in gold foil to simulate coins. Brooklyn-based Barton's, founded in 1938, made kosher chocolates for both Chanukah and Passover.

Valentine's Day (14 February), purportedly celebrated in honour of a saint killed in Roman times, was a popular holiday in medieval Europe. Precisely when candy became a Valentine's Day tradition is unclear, but in 1860 the British confectioner Richard Cadbury introduced the first Valentine's Day chocolate box, and couples still exchange elaborately boxed chocolates on that day. In America, Sweetheart candies, little sugar hearts bearing brief romantic mottos, were made by the New England Confectionery Company in 1902. By the twenty-first century, NECCO was producing about eight billion Sweethearts each year—virtually all of them sold during the six weeks before Valentine's Day.

Chocolates

Hot chocolate was consumed with sugar in Europe and North America by the mid-seventeenth century, but chocolate was not sold as a sweet until the nineteenth century. In 1815, Coenraad Van Houten, a Dutchman, developed a process for defatting chocolate and then subjecting it to an alkalizing process. This set in motion a series of discoveries that eventually made possible the manufacture of powdered cocoa, which was achieved in 1828. Eventually this led to the large-scale manufacture of chocolate in both powder and solid form.

Handmade chocolates were produced in England by the mid-nineteenth century. One manufacturer was John Cadbury, a Quaker and a strong temperance advocate who thought that it was important to provide alternatives to alcohol. In 1831, Cadbury began manufacturing cocoa for drinking chocolate; by 1866, Cadbury was also producing eating chocolate—handmade bonbons, chocolate-covered nougat and other chocolate candies. Cadbury began to produce milk chocolate in 1897. Another important chocolate manufacturer was Joseph Storrs Fry, another British Quaker, who invented a process for combining cocoa powder, sugar and melted cocoa butter to produce a thin paste that could be shaped in a mould to make chocolate bars. Soon, J. S. Fry & Sons was the largest manufacturer of chocolates in the world. In 1919, Cadbury acquired J. S. Fry & Sons, although a few Cadbury chocolate bars still bear the name 'Fry's'.

Cadbury itself was acquired by Kraft Foods in 2010, but

three years later, Kraft span off its sweets and snack food to a new company, Mondelēz International. Today, Mondelēz's bestselling confectionery brands are Cadbury's Dairy Milk, Milka chocolates and Trident gum. These rank sixth, fifth and third in global sales.

The grocers William Tuke and Sons of York, England, began to sell cocoa in 1785. In 1862, Henry Isaac Rowntree acquired the Tukes's cocoa business. In 1881, the company introduced Rowntree's Fruit Pastilles and in 1893, Rowntree's Fruit Gums. Four years later, Rowntree & Company was established. Like Cadbury, Rowntree innovated exceptional benefit programmes for employees, including dining and other facilities, workers' councils, a pension scheme, unemployment benefits and annual paid holidays.

In 1931, Rowntree began an aggressive development programme. One key to its success was the company's relationship with Forrest Mars, whose Mars Bar had been introduced in England in 1932. Until then, 'combination bars' (which had multiple ingredients such as chocolate, peanuts, caramel and so on) had not been popular in Britain. The Chocolate Crisp was launched in 1935 and renamed the Kit Kat two years later. The Rowntree company introduced Smarties in 1937. These colourful sugar-coated chocolate drops remain popular today in the United Kingdom, South Africa, Canada and Australia. Nestlé acquired Rowntree in 1988; subsequently many new chocolates and sweets have been introduced under the Rowntree brand.

American Chocolate Makers

Milton Hershey was a manufacturer of caramel candies in Lancaster, Pennsylvania. In 1893, he visited Chicago's Columbian Exposition and was fascinated by the chocolate-making machinery exhibited by Lehmann & Company of Dresden, Germany. Hershey bought Lehmann's machinery at the Exposition and had it shipped to Lancaster. Then he hired two chocolate makers from Baker's Chocolate and began mass-producing chocolate candies. Up to this time, all American chocolate candies had been made by hand. The Hershey Chocolate Company, a small subsidiary of Hershey's caramel business, initially produced breakfast cocoa, sweet chocolate, baking chocolate and a variety of small candies, eventually coming out with the Hershey's Milk Chocolate bar about 1905 and Hershey's Kisses in 1907. While the Hershey Chocolate Company was successful from the start, it received a major boost during the First World War, when Hershey chocolate bars were given to American soldiers fighting in Europe. Many had never eaten a chocolate bar before, and when they returned after the war, the demand for Hershey's products surged. Recently, the Hershey Company has expanded its global presence through increased sales and acquisitions. Its Reese's chocolates rank first in the U.S. and fourth globally in sales.

In 1922, Frank Mars of Minneapolis founded the Mar-O-Bar Company. It initially sold a bar composed of caramel, nuts and chocolate. The next year, the company introduced the Milky Way bar and in 1930, Snickers, a peanut-flavoured nougat bar topped with nuts and caramel and coated with chocolate. It quickly became one of the most popular candy bars in America, a position it has held ever since.

Frank Mars's son, Forrest Mars, did not get along with his father, so with $50,000 in his pocket he moved to England and formed a new company, Mars, Ltd. In 1932, he introduced the Mars bar, a slightly sweeter version of the Milky Way. By 1939, the company was ranked as Britain's third-largest confectionery manufacturer. When the Second World War started in 1939, Forrest Mars returned to the United States, where he launched a new company with Bruce Murrie, the son of the president of the Hershey Company. Because both of their last names started with M, they called their new company M&M. Their first product was a small milk-chocolate drop covered with a hard sugar shell, a small Smarties clone, which they named M&M's Chocolate Candies. The two companies, Mars and M&M, were not merged until 1964. Mars, Inc. has continued to expand, both through acquisitions and by creating new products. As of 2011, Mars controls 15 per cent of the global candy market and is the world's largest confectionery manufacturer. Their Dove chocolates (Galaxy in the UK), Orbit and Extra globally rank fifth, eighth and ninth respectively. For decades, Mars' M&M's were the world's largest selling confection. Sales hit $3.49 billion in globally in 2012, but M&M's were dethroned by Mars' Snickers, which had worldwide sales of $3.57 billion and is now the largest selling confectionery item in the world.

Other Confectionery Manufacturers

In the 1860s, Henri Nestlé, a German-born pharmacist living in Switzerland, developed condensed milk and successfully marketed an infant formula made from milk and flour. He sold his company in 1874, but it retained his name. Henri Nestlé then worked with his friend Daniel Peter, a chocolatier, to help perfect the milk chocolate bar. Peter's chocolate, made with Nestlé condensed milk, quickly became one of Europe's best-known chocolate brands.

Ovaltine, a Swiss product created by a physician in 1904 to nourish seriously ill patients, was widely marketed during the following decades. A sugary, malted chocolate powder designed to be mixed with milk and drunk hot or cold, Ovaltine's healthful qualities, particularly added vitamins, were touted in its advertising. Ovaltine's success encouraged Nestlé to market its own milk modifier: Nestlé Quik (also called Nesquik), a sweetened chocolate drink powder. It was introduced in 1948, and its long sponsorship of children's television programming ensured its lasting popularity. Its product line was subsequently extended to include sugary syrups and cereals.

The First World War depressed Nestlé's sales, but after the war the company began to expand. By the late 1920s, chocolate was its second most important product. In 1929,

Nestlé acquired Daniel Peter's company and entered into manufacturing powdered chocolate for making chocolate milk, premium chocolates and solid chocolate bars. After the Second World War, Nestlé began to grow rapidly, in part by acquiring other companies. In 1988, Nestlé acquired the Italian chocolate maker Perugina and the English chocolate maker Rowntree. Its Kit Kat bar ranks tenth in sales globally.

Another large confectionery manufacturer is Perfetti Van Melle Group, which was formed in 2001 and is headquartered in Milan, Italy. Its Mentos mints and chewy sweets rank eleventh in global sales today.

Tens of thousands of brands of assorted sweets are manufactured worldwide. Swedes consume more sweets than any other people in the world (annually 37 lb, or 16.8 kg, per capita). The Swiss consume more chocolate (annually 25 lb or 11.3 kg per capita). Americans eat less candy per capita, but spend more money—hitting $32 billion per year—and the amount is still increasing, even during the recent economic downturn. Worldwide sales of confectionery have also been increasing, and today are estimated at $150 billion annually.

Chapter 6
American Bliss

In addition to candies and confections, sugar is added to a large number of other products sold throughout the world, including breakfast cereals, biscuits (cookies), doughnuts, ice cream and soft drinks. Sugar is also added to processed foods that were not traditionally sweet, and where its flavour may not be dominant. Hidden sugar can be found in canned soups and vegetables; breads, crackers and chips; frozen dinners; condiments (ketchup, chilli sauce and Worcestershire sauce) and salad dressings; peanut butter; baby foods and infant formula; pizza; hot dogs and lunch meat; pickles and cocktail snacks; flavoured yogurt; frozen foods; fruit juice, fruit coolers, 'energy' and sports drinks; and even pet foods. Sugars hide in processed foods under a variety of names, including sucrose, glucose, dextrose, maltose, lactose, galactose, malt syrup, maltodextrin, corn syrup, high fructose corn syrup, molasses and corn sweetener, to name a few.

Nowhere is this sugarization of processed foods more apparent than in the United States. As sugar prices declined in the nineteenth century, sweet desserts and snacks became universal in American homes, regardless of income or social class. More and more sugar went into cakes, cookies, pies and other pastries. Foreign visitors remarked upon this, noting that the amount of sugar and other sweeteners 'used in families, otherwise plain and frugal, was astonishing'. By the 1870s, U.S. sugar consumption per capita was 41 lb (about 18.6 kg) per year; as commercially processed foods came on the market and sugar prices sank further, American consumption shot up. Cakes, from simple to lavish, were becoming an everyday part of the American diet. Parties were celebrated with a profusion of jelly cakes, pound cakes, plum cakes and lady cakes. From American kitchens issued a steady, fragrant stream of sugar cookies, wafers, kisses, drops, jumbles, snaps, macaroons, gingerbread, crullers and doughnuts—all with ever-greater quantities of sugar. Sweet rolls and doughnuts became regular breakfast fare. By 1901, Americans were consuming an average 61 lb (about 27.7 kg) of sugar per capita per year. The American love affair with sugar was flourishing, but was still a long way from its peak.

Breakfast Cereal

Until the twentieth century, the typical American breakfast included fruit, breadstuffs, eggs, potatoes and meats of all kinds—not just bacon or sausage but beefsteak, savoury meat pies and calves' liver. During the late nineteenth century, vegetarians and health reformers began to develop breakfast foods based on unrefined whole grains, which they thought better suited to the digestion of the modern-day office worker. The first commercial cereals were unsweetened and were meant to be moistened with plain water. As the industry took off, entrepreneurs found that customers preferred their cereal sweeter, so the new fashion was to serve it with cream and sugar. Will Kellogg added sugar to the formula for Corn Flakes over the objections of his brother, the health food guru and vegetarian John Harvey Kellogg, who believed that ingesting sugar carried more potential health risks than eating meat.

As more women entered the workforce during the twentieth century, cereals were advertised as a means of easing mother's workload. Children could prepare their own breakfasts without help and they loved sugary cereal. According to the medical authorities of the day, cereal was good for children, so this was a double win for busy mothers. Cereal companies, largely because they aimed their marketing directly at children, did well during the Depression; adding more and more sugar to their products helped seal the deal. After the Second World War, with sugar rationing a thing of the past, cereal makers upped the ante even further. In 1949, Post Cereals introduced Sugar Crisp, puffed wheat with a crunchy sugar coating. It was an immediate success, and other cereal companies followed suit with heavily sweetened cereals targeted at children. Some cereals approached 50 per cent sugar by weight; Kellogg's Honey Smacks hit 55.6 per cent sugar; Post responded with Super Orange Crisps, which weighed in at 70 per cent sugar. This led observers to ask, 'Is it cereal or candy?'

These high-sugar cereals were heavily promoted in children's media, especially radio and television, point-of-sale marketing and, later, the Internet. The big three American cereal manufacturers (Kellogg's, Quaker Oats and Post) spent more on advertising their products then they did on the ingredients that went into them. Annually the U.S. cereal industry uses 816 million lb of sugar, or almost 3 lb (about 1.4 kg) of sugar per capita. Ironically cold breakfast cereals, which started out as health foods, are now considered major contributors to excess sugar in the American diet, especially those of children. More than 1.3 million cereal commercials air on American television each year, and most of them are aimed at children.

Biscuits, Cookies, Cakes and Bread

The English word *biscuit* comes from Latin via Middle French; its original meaning was 'twice baked'. Some early European recipes (like those still used for Italian *biscotti*) called for dough to be baked in loaf form, then sliced or split and baked again, slowly, to drive off any moisture; the drier the product, the longer it would keep. British biscuit recipes came to America with the English colonists, but the Dutch also colonized parts of America and their word *koekje*, meaning small cake, became the American term for a sweet biscuit. Amelia Simmons, author of *American Cookery* (1796), is credited with publishing the first known 'cookie' recipes, including one for 'Christmas Cookeys' made with a pound and a half of sugar to three of flour.

Cookies are the simplest of baked goods and require few ingredients—sugar being one of them. They take only a few minutes in the oven, so they can be baked on short notice and served as an informal dessert or eaten as a snack. A full cookie jar was long seen as emblematic of a well-run American home, a loving mother and a happy family, and the cookies in the jar should be (it went without saying) home-baked: sugar-cookie cutouts, oatmeal or peanut-butter cookies or the classic chocolate-chip cookies. But neighbourhood bakeries and pastry shops offered more elaborate cookies, and commercial baking plants in the U.S. started to churn out mass-produced cookies in the nineteenth century. By the turn of the twentieth century, store-bought cookies were available nationwide, and advertising campaigns strove to make them acceptable to the fashionable hostess. The National Biscuit Company (later Nabisco), founded in 1898 as a conglomerate of smaller baking companies, pioneered a wide variety of cookies, such as Oreos, today the world's largest-selling cookie.

Cakes as we know them today began as a variation on breads. Some, like pancakes, were flat, and were turned to cook on both sides on a hot surface. Other cakes were baked in specially designed cake pans. If early cakes were sweetened, it was with a little honey, or they might be baked unsweetened and served with honey as an accompaniment.

During the sixteenth and seventeenth centuries, sugar replaced honey in cake batters, and sugar-based icings or frostings supplanted the accompanying honey. By 1615, cookbooks were advising the use of 'a good deale of sugar' in cake recipes. By the 1680s, cakes were commonly served as a dessert after a meal, or with tea or coffee in the morning or afternoon. Lavishly decorated cakes became a feature of special occasions and ceremonial feasts, such as Christmas, weddings and birthdays. As sugar prices declined and sugar refining improved, the amount of refined sugar used in cakes increased; powdered or icing sugar (also called confectioners' sugar because it was used in candy- making) became widely available in the nineteenth century, when it began to be called for in recipes for cake icing.

Cake-baking traditions were brought to America by European immigrants, and cakes were popular in colonial times; they remain one of America's favourite desserts. From simple gingerbread, pound, angel food and sponge cake to rich fruitcakes, cheesecakes, frosted and filled layer cakes, elaborate wedding cakes and whimsically decorated cupcakes, American cakes call for generous amounts of sugar. Although many people still bake cakes from scratch at home—especially birthday cakes—there is an abundance of mixes, bakery cakes and packaged or frozen products for those not so inclined.

Historically bread was made without the addition of sugar, although some early nineteenth-century recipes published for brown bread (also called dyspepsia bread) included molasses. This changed when millstones in flour mills were replaced by high-speed steel rollers during the late nineteenth century. The bran, germ and oil were removed from the wheat to produce a bland white flour. Bakers began to compensate for this tastelessness by adding sugar, and the amount of added sugar increased over time (sugar also adds moisture to bread, so it stays fresh longer). In the 1880s, cookbook authors recommended one tablespoon of sugar to every eight cups of flour. By the 1890s, this had increased in some recipes to about one tablespoon per cup of flour. Commercial bakers added even more. This increased even more for commercial bread during the twentieth century. By comparison, bakers in other countries, such as Italy and France, include little or no sugar in their bread.

Doughnuts

American doughnuts (or donuts) may have been of Dutch, German or English origin. The Dutch called them *olijkoeken* (oil cakes) or *oliebollen*. These were pinched-off portions of sweetened dough that were rolled between the hands and then dropped into hot oil. The Dutch-style nuggets or 'nuts' of fried dough were popular in America, but doughnuts with holes in the centre were not common until the end of the nineteenth century. Purportedly the hole was a practical innovation that made for easier dunking in coffee. Others maintain that the shape helped the dough to cook more evenly.

The sale of commercial doughnuts greatly expanded after the Second World War. Doughnut retailing lends itself to franchising because the equipment is affordable. Doughnut franchisers include Dunkin' Donuts, House of Donuts, Krispy Kreme and Winchell's. Dunkin' Donuts alone sells an estimated 6.4 million doughnuts per day (2.3 billion per year).

Yet another doughnut chain was established by Tim Horton, a hall-of-fame Canadian hockey player, who opened his first outlet in Hamilton, Ontario, in 1964. It was known for its coffee, which was guaranteed to be served fresh, cappuccinos, doughnuts and 'donut holes', but it quickly incorporated other items. The company soon expanded and it became Canada's largest fast food operation. In 1995, Tim Hortons opened outlets in the United States. When Burger King agreed to purchase the chain in 2014, Tim Hortons had almost 4,600 systemwide restaurants, inluding about 845 in the United States and other countries.

About 80 per cent of doughnut business is take-out, and 80 per cent of doughnuts are sold before noon in North America. They come in a great diversity of shapes, sizes and flavours. There are yeast-raised doughnuts and baking-powder doughnuts, both deep-fried; for the fat-avoidant, oven-baked doughnuts are available. Most doughnuts have holes, and doughnut holes (or pieces of dough shaped to resemble them) are sold separately. Filled doughnuts are injected with jam (jelly), custard or a variety of other sweet fillings; further adornments may include a thick dusting of icing (confectioners') sugar or cinnamon sugar, a thin layer of chocolate icing, a vanilla, chocolate or other flavour glaze, toasted coconut or sprinkles. Similar pastries are crullers (or krullers), large strips of dough twisted together and fried; and bismarcks, large, éclair-shaped jam doughnuts.

Ice Cream

Ices, ice creams, and sorbets—frozen desserts initially sweetened with fruit juice—likely originated in Italy or France in the sixteenth century. They were sold in cafés in Europe from the seventeenth century on, and small vendors operated in most European cities by the 1800s. Several ice-cream recipes appeared in English cookbooks in the eighteenth century. European immigrants brought ice-cream-making techniques to America, where ice-cream parlours had opened in some cities by the 1790s. Many ice-cream recipes—most sweetened with plenty of sugar—appear in nineteenth-century American cookbooks.

Three main ice-cream flavours—chocolate, vanilla and strawberry—came to the fore in the nineteenth century and have remained favourites ever since. But other flavours proliferated during the nineteenth century, as did elaborate ice-cream presentations with inventive toppings, sauces and garnishes, and soda-fountain drinks made with ice cream. The late twentieth century saw the rise of the mix-in—

premium ice cream studded with chunks of cookies, candies, chocolate, nuts or fruit, or with thick swirls of caramel, fudge or peanut butter running through the cream.

For much of the nineteenth century, going out for ice cream was a genteel pursuit. Served in establishments called 'parlours', ice cream was scooped into elegant glass dishes and eaten with a spoon. It was a popular summer treat, and the ingredients were cheap, but the problems for street vendors were how to keep the product cold and how to serve it without dishes and spoons. The solution was the ice cream cone, which was invented in the late nineteenth century.

Commercial production of ice cream did not take hold until technological improvements in refrigeration made sales possible through drugstores, soda fountains and grocery stores. In the United States, soda fountain offerings competed with the alcohol served in saloons and bars, and were therefore championed by the temperance movement. Prohibition gave a huge boost to the popularity of ice cream as bars, saloons and taverns were shuttered and soda fountains became community gathering places. However, it was not until after the Second World War, when self-serve freezer chests for grocery stores came into wide use and the freezer sections of home refrigerators increased in size and efficiency, that packaged ice cream became an everyday part of the diet.

By the 1950s, large ice-cream makers were underselling small producers, and supermarkets switched to national brands. But a niche had opened up for 'super-premium' ice creams, with a higher butterfat content and less air than supermarket brands. Häagen-Dazs, a new product from a decades-old family ice-cream business, first appeared in 1960, and Ben & Jerry's ice cream, originally made by hand by two young men from Long Island, was first sold from a converted gas station in Burlington, Vermont, in 1978. Breyers is still the largest ice-cream manufacturer in America, a position it has held since 1951. It is followed by Dreyer's/Edy's and Blue Bell Creameries, Inc. Despite the concentration of the ice cream industry, the largest category of ice-cream makers in America today is private labels, generally sold at the local and regional level. In 2013, Americans purchased an estimated $11 billion worth of ice-cream—complete with a generous helping of sugar in virtually every serving.

Sugary Beverages

Yet another sugar-filled treat is soda pop, which, like breakfast cereal, began life as a health food and ended up as just the opposite. Mineral waters, both still and naturally effervescent, have long been considered therapeutic, and water artificially infused with carbon dioxide (CO_2) was considered to have medicinal attributes. At European spas and resorts built at natural springs, drinking bubbly mineral waters was an important part of the health regimen. During the eighteenth century, several scientists, including Joseph Priestley and Antoine-Laurent Lavoisier, discovered that carbon dioxide was the source of the bubbles in natural springs, beer and champagne. Priestley constructed an apparatus for manufacturing the gas, and reports of his invention were sent to John Montagu, the fourth Earl of Sandwich (the same man credited with inventing the sandwich), who was then Lord of the Admiralty. He requested that Priestley demonstrate his apparatus before the Royal College of Physicians. Priestley did so; among the audience members was Benjamin Franklin, who was living in London at the time.

Other scientists constructed their own systems for producing soda water. In 1783, Johann Jacob Schweppe improved a process for manufacturing carbonated water and formed the Schweppes Company in Geneva, Switzerland. During the French Revolution and its aftermath, Schweppes moved his operation to England, where his soda water was approved for medicinal use by the British royal family.

By 1800, manufacturers had found they could make water fizzy by adding a solution of sodium bicarbonate to it. Carbonated water, however, was generally made under high pressure using sulphuric acid. Operators could easily be burned by the acid, and containers sometimes exploded. Various kinds of apparatus for making carbonated water were patented from 1810, but because of the complexity of the process, they could be operated only by trained technicians. Because the devices were expensive, and the beverages made with them were considered medicinal, soda water was generally dispensed only in drugstores. It was a small step from carbonated water to flavoured soda water. Ginger ale is generally thought to have been the first flavoured carbonated beverage sold commercially in America. It was probably first marketed in 1866 by James Vernor, a Detroit pharmacist, who created Vernors Ginger Ale.

Another early soft drink was root beer, which was traditionally flavoured with bark, leaves, roots, herbs, spices and other aromatic parts of plants. In its early years, root beer was a home-brewed, mildly alcoholic beverage. Later, extracts made from the flavourful ingredients were touted as a tonic—typical of herbal remedies of the period. By the 1840s, root beer mixes and syrups were manufactured locally and sold in confectionery and general stores. Soda fountains, which sold combinations of ice cream and drinks composed of fruit syrups, sugar and soda water, sprang up around America.

Soda companies produced a sugary syrup or extract and sold it to drugstores, where it would be combined with carbonated water. This began to change in 1892, when William Painter invented the crown bottle cap, which made it possible to seal bottles easily, cheaply and securely. At the same time, bottling technology improved: the new, stronger glass bottles could hold the 'fizz' without shattering during bottling.

Soft drinks got another major boost during Prohibition, when manufacturing and selling alcoholic beverages was illegal. It was also during the 1920s that fast food chains emerged, and virtually all of them sold soft drinks. When Prohibition was repealed in 1933, soft drinks and fast food outlets were already well-established American institutions and they continued on their upward sales trajectory.

Soft drink manufacturers spend billions of dollars on promotion and advertising. Marketing efforts are aimed at children through cartoons, movies, videos, charities and amusement parks. In addition, soft drink companies sponsor contests, sweepstakes and games via broadcast and print media as well as the Internet, much of it targeting young people. In its study *Liquid Candy* of 2005, the Center for Science in the Public Interest (CSPI) revealed that soft drink companies had targeted schools for their advertising and sales of their products. It also reported that soft drinks 'provided more than one-third of all refined sugars in the diet'. Soft drinks, according to CSPI, are the single greatest source of refined sugar, providing 9 per cent of calories for boys and 8 per cent for girls. The CSPI study also reported that at least 75 per cent of American teenagers drink soda every day.

American soda companies have rapidly expanded abroad. Coca-Cola and PepsiCo sell more than 70 per cent of the carbonated beverages in the world. Worldwide, soda companies sell the equivalent of 1.3 billion glasses of soda every day, which works out to about eight teaspoons of sugar per glass of non-diet soda.

Energy and Sports Drinks

Added sugar is also found in many other beverages,

including fruit juices, fruit coolers, coffee beverages and 'energy drinks'. Sugar has been added to fruit drinks ever since they emerged as a mass-consumption processed product in the 1940s.

Manufacturers have used the word 'fruit' in their beverage names to persuade potential buyers of good nutrition inside the can or bottle, but many fruit drinks are just fruit-flavoured sugar water. Fruit coolers, for instance, typically have 16 grams of sugar. Other beverages have more. A 20 oz (600 ml) bottle of Vitamin Water contains 33 grams of sugar. A 16 oz (475 ml) Starbucks Café Vanilla Frappuccino has 67 grams per serving. A Costa Medio Tropical Fruit Cooler contains 73 grams of sugar—seven times more than a Krispy Kreme doughnut.

Energy drinks have become pervasive and most are filled with sweeteners. An English pharmaceutical company developed the first energy drink, called Glucozade, in 1927. It was a fizzy liquid filled with sugar, and was mainly used to help children recover from illness. A British pharmaceutical picked up the formula, renamed it Lucozade, and promoted it with the slogan 'Lucozade aids recovery.' In 1983, the company decided to reposition the product as an energy drink using the slogan, 'Lucozade replaces lost energy.'

Red Bull, consisting of caffeine, sucrose, glucose and other ingredients, was launched in Europe in 1987. It was introduced into the United States a decade later, when it became America's first popular energy drink. Red Bull started a frenzy of copycat beverages, such as Jolt, Monster

Energy, No Fear, Rockstar, Full Throttle and a myriad of other brands. Large companies jumped in: Anheuser-Busch's 180, Coca-Cola's KMX, Del Monte Foods' Bloom Energy and PepsiCo's Adrenaline Rush. These drinks boast caffeine levels of up to 500 mg per 16 oz (475 ml), and they are often loaded with various forms of sugar. By the early twenty-first century there were more than 300 branded energy drinks on the market in America alone—and most were filled with high-calorie sweeteners.

Sugar is also a major ingredient in many sports drinks—beverages that are designed to enhance athletic performance by fostering endurance and recovery. The first such beverage was Gatorade, formulated in 1965 by Robert Cade and Dana Shires of the University of Florida. Gatorade is a noncarbonated drink that consists of water, electrolytes and a heavy dose of carbohydrates (in this case, mainly sugar—28 grams of which are in 16 oz of the drink). Gatorade launched the sports-beverage industry. Intended for athletes participating in serious competition or intense exercise, sports drinks do increase energy levels (as do all sugar-sweetened drinks)—but most sports and energy drinks are consumed by non-athletes, who often end up gaining weight.

Bliss Point

That sweetened products sell well has been known since the early twentieth century. Manufacturers wanted to know how much sugar should be added for maximum sales. The research started in the early 1970s, when the psychologists Anthony Sclafani and Deleri Springer engaged in an experiment to induce obesity in laboratory rats. They found that rats did not overeat or gain weight when they were fed only Purina Dog Chow, but became obese when given Froot Loops, a high-sugar breakfast cereal. Sclafani and Springer repeated their experiment using other common supermarket foods—peanut butter, marshmallows, chocolate bars, sweetened condensed milk and chocolate-chip cookies. The rats preferred sweet food—and, if given the opportunity, continued to eat it until they became obese. Other experiments subsequently proved that when obese rats were exposed to non-sweet standard rat food, they declined to eat it.

About the same time, Howard Moskowitz, a researcher in the U.S. Army labs in Natick, Massachusetts, was searching for ways to make military rations more palatable for soldiers in combat. His experiments showed that soldiers' preferences for foods increased as sugar was added—up to a point—but beyond that point, additional sugar made the food *less* appealing. Moskowitz is credited with coining the term 'bliss point' to describe that peak in the appeal of sweetness (bliss points were also established for fat and salt intake). The conclusion of these and other studies was that a love for sugar was inborn, and that humans were hardwired to prefer sweet foods. In 1981, Moskowitz left the Army labs and opened his own consulting firm in White Plains, New York, where many American food companies were headquartered. His company helped food corporations find the 'bliss point' for their products. He was extremely successful and so were the companies he advised.

Yet another set of experiments was conducted at the Monell Chemical Senses Center in Philadelphia, an independent non-profit research facility funded by governmental agencies and large corporations. Researchers concluded that children, in particular, preferred sweeter foods than adults. Later experiments at the Center found that a preference for sweet flavours was a basic part of children's biology, and allowed researchers to determine the exact bliss point for sugar in children's foods and beverages. Further studies around the world, such as at the London-based—and corporate-funded—ARISE (Associates for Research into the Science of Enjoyment), confirmed these studies and concluded that the taste for sweetness was inborn.

These studies helped food manufacturers figure out how much sugar to incorporate in their products in order to stimulate sales. Candy and cereal companies, bakers and soft-drink manufacturers throughout the world raised the levels of sweetness in their products to the scientifically identified bliss point. The consumption of sugary foods and beverages rapidly expanded—as did the waistlines of consumers around the world, generating considerable criticism for the processed and fast food industries.

Chapter 7
Sugar Blues

Concerns about the health effects of consuming sugar have been expressed for the past four centuries. The main early concern was the relationship between sugar consumption and dental caries (cavities). One of the first references to this is in the writings of Paul Hentzner, a German visitor to England who met the 66-year-old Elizabeth I in 1598. He described her as having black teeth, commenting that it was 'a defect the English seem subject to from their too great use of sugar'. The medical authorities agreed: sugar rotted the teeth. William Vaughan, a doctor of civil law, condemned sugar for various reasons; one, as he wrote in his work *Approved Directions for Health* (1612), was that it blackened and 'corrupted' teeth. James Hart, author of *Klinike; or, the Diet of the Diseases* (1633), proclaimed that the immoderate use of sugar in sugar candy, sweet confections and sugar-plums produced 'dangerous effect in the body,' including constipation, consumption, blockages and 'rotten teeth, making them look blacke'. He went on to warn 'young people especially to beware' of consuming these confections. Writers continued to note the relationship for centuries.

Jonathan Swift, for instance, proclaimed that 'sweet things are bad for the teeth' in one of the dialogues in *A Complete Collection of Polite and Ingenious Conversation* (1722).

American medical authorities later expressed their own concerns about refined sugar. The health advocate Sylvester Graham, in his final work, *Lectures on the Science of Human Life* (1839), advocated banning refined sugar because it was stimulating: 'The stern truth is, that no purely stimulating substances of any kind can be habitually used by man, without injury to the whole nature.' Many health reformers followed Graham's beliefs. The hydropath Russell Trall violently attacked sugar in editorials and in his books on health:

> Sugar is made into an immense variety of candies, confections, lozenges, etc., most of which are poisoned with coloring matters, and many of which are drugged with apothecary stuff. The intelligent physiologist will repudiate their employment in every form or shape. The raw sugars of commerce contain various impurities; and the refined and very dry sugars tend to constipate the bowels.

Not all health reformers heeded Trall's absolutist views on sugar. John Harvey Kellogg, the Seventh-Day Adventist who directed the sanatorium in Battle Creek, Michigan, felt that his childhood gastrointestinal problems were caused by meat and candy, and believed that the American love of candy and sweet desserts needed to be rigidly controlled, for sugar interfered with proper digestion. But he did not call for the complete elimination of sugar, recommending only that people should eat much less of it, and replace it with honey, dates and raisins.

Dentists, understandably, condemned sugar, but medical authorities were also concerned. In 1942, the American Medical Association's Council on Food and Nutrition stated that 'it would be in the interest of the public health for all practical means to be taken to limit consumption of sugar in any form in which it fails to be combined with significant proportions of other foods of high nutritive quality.' The medical profession also worried about sugar consumption, particularly its effects on hypoglycaemia (low levels of sugar in the blood). E. M. Abrahamson, MD, and A. W. Pezet, in their book *Body, Mind, and Sugar* (1951), concluded that refined sugar caused a 'constellation of diseases', and that its removal from the diet led to immediate improvement in patients' health. The book was mainly based on personal experiences—Abrahamson was a physician specializing in diabetes who dosed Pezet with 'hyperinsulinism'—but it received wide publicity and sold more than 200,000 copies. Other medical professionals agreed. A British Royal Navy surgeon, Thomas L. Cleave, and a South African physician, George D. Campbell, examined a number of different societies and found that diabetes, heart disease, obesity, peptic ulcers and other chronic diseases were correlated with increased consumption of refined sugar, white flour and white rice. The less refined carbohydrates that were consumed, the lower the incidence of these diseases. They published their views in the book *Diabetes, Coronary Thrombosis and the Saccharine Diseases* (1966). While their beliefs were scoffed at by some, most medical professionals did recommend that their patients reduce their intake of added sugar.

Sugar Substitutes

Concern with increasing diabetes and obesity led to the invention of non- caloric and low-calorie artificial sweeteners. The first artificial sweetener was saccharin, a white crystalline powder that has 300 to 500 times the sweetness of sugar, but no calories. It was discovered by a graduate student at Johns Hopkins University in Baltimore in 1879. Saccharin was commercialized by several companies, including Monsanto, but was not widely used until sugar rationing was encouraged during the First World War.

After the war, saccharin became a boon to diabetics, and eventually was used in diet products for weight-loss regimes. In 1977, a Canadian study reported that saccharin caused cancer in test animals, and the U.S. Food and Drug Administration (FDA) placed a moratorium on its use until more studies were conducted. Further studies did not confirm earlier test results and the ban was lifted in 1991.

A second artificial sweetener, calcium cyclamate, was

used in diet soda beginning in 1952. Variations of it were used in a variety of other products. Lab studies carried out in the late 1960s showed that cyclamates were likely carcinogenic, and they were banned by the FDA in 1970. By the late 1970s, most diet products were being made with a third sugar substitute, aspartame (marketed under the names NutraSweet and Equal).

Stevia is a non-caloric natural sugar substitute derived from plants in the sunflower family. The plant extract is 300 times as sweet as table sugar. Stevia became popular in Japan in the 1970s and since then has been widely used in many countries in Asia and South America. In 1994, the FDA classified stevia as a herbal supplement and required it to be listed on the food labels. In 2008, the FDA approved two sweeteners derived from stevia: Truvia, developed by Cargill and the Coca-Cola Co., and PureVia, developed by PepsiCo and the Whole Earth Sweetener Company. The following year, the FDA placed a purified form of stevia on the 'Generally Recognized as Safe' list.

Another recent introduction is Sucralose, which is 600 times sweeter than table sugar. It was approved for use in Canada in 1991, and seven years later in the U.S. Marketed under several brand names—Splenda, SucraPlus, Candys, Cukren and Nevella—it is found in thousands of diet products. Acesulfame potassium, another non-caloric sweetener, is 200 times as sweet as table sugar. It is approved for use in the United States and the European Union. Neotame, an artificial sweetener made by NutraSweet, is 7,000 to 13,000 times sweeter than table sugar. The FDA approved it for use in the U.S. in 2002, but it has not yet come to be widely used. Little evidence has surfaced about negative short-term effects of approved artificial sweeteners; the long-term health risks, if any, remain under debate.

Empty Calories

The term 'empty calories', meaning calories derived from foods with few or no nutrients other than carbohydrates or fat, was first employed during the 1950s. At the top of the empty calorie food chart are sugary foods and drinks such as candy, cookies, cakes, pies, ice cream, breakfast cereals and sodas.

John Yudkin, who established the nutrition department at the University of London in 1953, was convinced that there was a clear connection between sugar consumption and many chronic diseases. By the end of the 1950s, Yudkin was campaigning for the elimination of sugar from the diet to prevent coronary heart disease and aid in weight loss. In 1958, he published a diet manual, *The Slimming Business* (1958), which advocated a very low-carbohydrate diet for weight loss. Yudkin published a number of studies supporting his views, which were popular in Britain during the 1960s. In 1972, he published *Pure, White and Deadly: The Problem of Sugar*, a tirade that generated considerable interest among the public in Britain and the United States; still, his views were generally rejected by the medical community, which concluded that dietary fat—not sugar—was the major cause of heart disease.

Medical professionals in the United States concluded that American children consumed too much sugared baby food and sweetened breakfast cereals, resulting in hyperactivity and other health problems throughout their lives. William Duffy, an American journalist and macrobiotic advocate, published the best-selling *Sugar Blues* (1975) about 'the multiple physical and mental miseries caused by human consumption of refined sucrose'. He compared sugar to heroin and called it at least as addictive as nicotine and just as poisonous.

Despite these warnings, Americans continued to increase their per capita sugar consumption, although not all of it was from sucrose. Much of it was from the misnamed high-fructose corn syrup (HFCS). In the 1950s, scientists had learned to refine corn into starch, then convert the starch into glucose, and finally convert the glucose into fructose by adding enzymes. Although it is made from corn, commercial HFCS is chemically similar to sucrose. HFCS contains 45 per cent glucose and 55 per cent fructose, while sucrose contains equal amounts of glucose and fructose. HFCS's advantage was that it was sweeter than sucrose. Its disadvantage was that it was more expensive than sucrose at the time. This changed in the 1970s, when the price of sugar increased in the United States due to quotas and tariffs on imported sugar and subsidies for corn growers which lowered the price for corn. American manufacturers added HFCS to products, particularly beverages. Most subsequent studies have concluded that the human body handles HFCS in the exact same way that it does sucrose. The majority of researchers now conclude that the health issues are associated with the total consumption of refined sugar, not with HFCS.

The term 'junk food', meaning calorie-dense processed foods, particularly sweets, salty snacks and fast foods and sugary beverages that have little nutritional value other than calories, was first used in the 1970s, It was popularized during the following decade by Michael Jacobson, the director of the Center for Science in the Public Interest (CSPI), which from its earliest days has decried high-sugar foods. The problem, according to the CSPI (and many others), is not just eating junk foods, but allowing them to crowd out more nutritious ones.

Refined sugar is an important contributor to excess calorie consumption. According to studies published in 2011 in the prestigious British medical journal *The Lancet*, the global prevalence of obesity has almost doubled since 1980, 'when 4.8 percent of men and 7.9 percent of women were obese. In 2008, 9.8 percent of men and 13.8 percent of women in the world were obese.' An estimated 1.3 billion people worldwide are overweight—half of whom are obese—and the number of people who are overweight is increasing in nearly every country in the world. Excess weight has been linked with high blood pressure, arthritis, infertility, heart disease, stroke, Type 2 diabetes, birth defects, gallbladder disease, gout, impaired immune function, liver disease, osteoarthritis and several types of cancer (including breast, prostate, oesophageal, colorectal, endometrial and kidney cancer).

There are many causes of obesity and overweight, but an examination of research findings led the research institute of the Zurich-based financial firm Credit Suisse to conclude in 2013 that 'While medical research is yet to prove conclusively that sugar is the leading cause of obesity, diabetes type II and metabolic syndrome, the balance of recent medical research studies are coalescing around this conclusion.' They believe that sugar meets 'the criteria for being a potentially addictive substance'. Nowhere is this problem more acute than in the United States, where 61 per cent of Americans are classified as overweight. The annual cost, concludes the Credit Suisse's Research Institute, is also staggering: 30 to 40 per cent of all healthcare expenditures in the United States—about $1 trillion—'go to help address issues that are closely tied to the excess consumption of sugar', it reports.

Epilogue

Sugar production and consumption remain easy targets for environmental, political and nutrition groups, and for very good reason. Environmentalists hold sugar cane growers responsible for the destruction of rainforests in Brazil, the degradation of the Great Barrier Reef in Australia and the deterioration of the Florida Everglades. Fertilizer and pesticide runoff have caused environmental damage throughout areas where cane or sugar beet are cultivated, including the pollution of fresh and oceanic waters. Contract labourers in sugar-cane-growing areas, such as Haitians in the Dominican Republic, are treated abysmally, and concerns have been raised about migrant workers employed in cane fields in many other countries, including the United States.

Citizens' groups have charged large agricultural interests and food corporations with lobbying for subsidies for domestic sugar production, and high tariffs and low quotas for imported sugar, such as those currently in place in the EU and the U.S. These actions have lowered the price of sugar worldwide, causing severe economic crises in some of the world's least developed countries, and have pushed up the cost of sugar-containing processed foods in developed countries.

Health professionals and nutrition experts have identified added sugar as a major cause of obesity and have pegged it as a contributing factor in many illnesses, including diabetes, heart disease, obesity, peptic ulcers and other chronic diseases. Critics blame large food corporations for lacing their products with excessive amounts of sugar, and for targeting children in their advertising campaigns on television, radio and the Internet, and at or near schools and sporting events.

Because of the criticism, some food companies have begun to decrease the amount of sugar in their foods. Since 2007 the Kellogg Company and General Mills, which produced six of the ten unhealthiest cereals, and did the most child-focused marketing of any cereal company, have lowered the sugar content of the cereals they market to children.

Despite the anti-sugar movement, sucrose remains one of the world's most important foods: it is estimated that about 8 per cent of the total calories consumed in the world comes from sugar, although the amount consumed by populations varies greatly. The global average is about 17 teaspoons (70 grams) per day. At the top are Americans, who consume an average 40 teaspoons of sugar per day—132 lb (about 60 kg) per year. Close behind are Brazilians, Argentinians, Mexicans and Australians, with 30 teaspoons per person, per day. Indians eat far less, but the Chinese consume the least among the world's largest countries at seven teaspoons per day—about 4 lb (about 1.8 kg) per year.

Sugar cane and sugar beet remain among the world's most important crops. Although they are grown in many countries, the major production breaks down to relatively few. Brazil is by far the world's largest grower of sugar cane, producing around 28 per cent of the global crop, but about half of it is converted into ethanol. Brazil exports about 25 per cent of the world's total processed raw sugar. India ranks second in production, and together with China and Thailand accounts for about a third of total world sugar production. The remainder is produced by 114 other countries around the world.

Sugar will remain an important part of the human diet. It isn't just our physiological needs, or the successful marketing of junk-food and soda manufacturers, that attract us to sweet foods and beverages. Sweet-tasting foods—candy, cake, chocolate, ice cream and soda—generate good feelings, and serve as small rewards that help us get through the day. They are often associated with good times—holidays and celebrations such as Christmas, Easter, Valentine's Day, Halloween, birthday parties and weddings. Consumed in moderation, sweet foods and beverages will remain an integral part of our lives far into the future.

Recipes

To make a March-pane

from *Delightes for Ladies* (London, 1611)

Take two pounds of almonds being blanched and dryed in a sieve over the fire, beate them in a stone mortar, and when they bee small, mixe them with two pounds of sugar beeing finely beaten, adding two or three spoonefuls of rose-water, and that will keep your almonds from oiling: when your paste is beaten fine, drive it thin with a rowling pin, and so lay it on a bottom of wafers; then raise up a little edge on the side, and so bake it; then yce [ice] it with rose water and sugar, then put it into the oven againe, and when you see your yce is risen up and drie, then take it out of the oven and garnish it with pretie conceipts, as birdes and beasts being cast out of standing-moldes. Sticke long comfits upright into it, cast bisket and carrowaies in it, and so serve it; you may also print of this marchpane paste in your moldes for banqueting dishes. And of this paste our comfit-makers at this day make their letters, knots, armes, escutcheons, beasts, birds, and other fancies.

Chocolate Cream

from François Massialot, *Le Cuisinier royal et bourgeois* (Paris, 1693)

Take a Quart of Milk with a quarter of a Pound of Sugar, and boil them together for a quarter of an Hour: Then put one beaten Yolk of an Egg into the Cream, and let it have three or four Walms: Take it off from the Fire, and mix it with some Chocolate, till the Cream has assum'd its colour. Afterwards you may give it three or four Walms more upon the Fire, and having strain'd it through a Sieve, dress it at pleasure.

To make Liquorish Cakes

from *The Pastry-cook's Vade-mecum; or, a Pocket-companion for Cooks, Housekeepers, Country Gentlewomen, &c.* (London, 1705)

Take 12 ounces of Liquorish scraped very thin, then take two pints and a half of Isop Water, one pint and half of Coltsfoot Water, a pint and half of red Rosewater, two good handful of Rosemary flowers, one handful of Maiden-hair, keep all these together three or four days in a stew Pot or Jug that may be close stop'd, shaking them together two or three times a day, then put them all into a Skilet, and set them upon a soft Fire two hours, then strain it into a Silver Bason, put to it a pound of brown Sugar-candy so let it boil till it grow thick enough to beat to a Paste, when you find it grow pretty thick, take a little upon a Spoon, and beat it with a Knife till it be cold, and then you will find whether it be enough, when you take it off the Fire, it must be beaten with a good strength with a Spoon till it be white, then take some fine Sugare searced, and so roul it up in little Cakes, the best way is to keep beating it to the last, or else it will so crackle that it will never role handsomely, half this Receipt is enough to make at a time.

Portugall Cakes

from Edward Kidder, *Receipts of Pastry for the Use of his Scholars* (London, np, c. 1720)

Put a pd [pound] of fine sugar & a pd of fresh butter 5 eggs & a little beaten mace into a flatt pan beat it up wth yor hands till tis very leight & looks curdling yn put thereto a pd of flower ½ a pd of currants very clean pickt & dryd beat yn together fill yor hart pan & bake ym in a slack oven

You may make seed cakes ye same way only put carraway seeds instead of currants.

Another Christmas Cookey

from Amelia Simmons, *American Cookery, 2nd edn* (Albany, NY, 1796)

To three pound flour, sprinkle a tea cup of fine powdered coriander seed, rub in one pound of butter, and one and half pound sugar, dissolve three tea spoonfuls of pearlash in a tea cup of milk, kneed all together well, roll three quarters of an inch thick, and cut or stamp into shape and size you please, bake slowly fifteen or twenty minutes; tho' hard and dry at first, if put into an earthern pot, and dry cellar, or damp room, they will be finer, softer and better when six months old.

To Clarify Sugar

from Hannah Glasse and Maria Wilson, *The Complete Confectioner; or, Housekeeper's Guide* (London, 1800)

In proportion to three pounds of fine, lump, or powder sugar, which you are to put in a skillet or boiler; break into an earthen pan the white of an egg, with near a pint of fresh water, and beat them up all together with a wisk to a white froth; then put the whole into a copper kettle, or pan, and set them on a clear and slow fire; when it begins to boil, do not fail to put a little more water in, and begin to skim it, till you see the scum appears thick on the top, and the sugar becomes pretty clear; that done, to clear it properly, sift it in a wet napkin, or silk sieve, and pass it thus into what vessel you please, till you want to make use of it.

Note.—If the sugar does not appear very fine, you must boil it again before you strain it; otherwise, in boiling it to a height, it will rise over the pan.

German Biscuits

from William Alexis Jarrin, *The Italian Confectioner* (London, 1829)

Take cloves, cinnamon, corianders, nutmeg, of each a quarter of an ounce, and pound and sift them (or the essence of those spices will answer the same purpose); two ounces of preserved lemon peel, and one pound of sweet almonds cut into fine prawlings [as for pralines]; mix these ingredients with twenty four eggs, and five pounds of sugar, and as much flour as will make it of a malleable paste. Roll it out into squares, lozenges, ovals, or any other shape; when baked put on them an iceing of chocolate &c. to your taste.

Of Boiling the Sugar

from M. A. Carême, ed. John Porter, *The Royal Parisian Pastrycook and Confectioner from the Original of M. A. Carême* (London, 1834)

There are six degrees of boiling the sugar after it is clarified, viz.; –

First Degree.—*Au Lisse.*—The sugar being clarified, put it on the fire, and after boiling a few moments, take a little of it on the top of your fore- finger, which you press against your thumb; when, on separating them immediately, the sugar forms a fine thread hardly visible, but which you can draw out a little, it is a sign that your sugar is boiled *au grand lisse* but if on the contrary it breaks instantly, your sugar is only *au petit lisse.*

Second Degree.—*Au Perle* (to a Pearl or Bead).—Having boiled your sugar a little longer than stated in the preceding degree, again take some between your fingers, which on separating them immediately will cause the sugar to extend like a string. When this string breaks, your sugar is called *au petit perle*, but if it extends from one finger to the other, without breaking, it is a proof that your sugar is boiled *au grand perle*.

The bubbles thrown up by the sugar in the latter case, should, besides, appear on the surface like small close pearls.

Third Degree.—*Au souffle*.—Continue boiling your sugar, and then dip a skimmer in it which you strike immediately on the pan. Then blow through the skimmer, and if that causes small bubbles to pass through it, it is a sign that your sugar is boiled *au souffle*.

Fourth Degree.—*À la Plume* (to a Feather).—Let the sugar boil up again; then dip in the skimmer and shake it hard, in order to get off the sugar, which will immediately separate itself from it, and form a kind of flying flax. This is called *à la grande Plume*.

Fifth Degree.—*Au Cassé* (to a Crack).—After boiling the sugar a little longer, dip the end of your finger first in cold water, and then in the sugar, and immediately after again in cold water, which will cause the sugar to come off your finger. If it then breaks short, it is boiled *au cassé*; but if, on putting between your teeth, it should stick to them, it is only boiled *au petit cassé*.

Sixth Degree.—*Au Caramel*.—When the sugar has been

boiled to the 5th degree, it passes rapidly to a caramel; that is, it soon loses its whiteness, and begins to be very lightly coloured, which proves that your sugar is really boiled to a caramel.

Sugar Candy

from L.-J. Blachette, *A Manual of the Art of Making and Refining Sugar from Beets* (Boston, MA, 1836)

To fulfil, without omission of anything, the task we have imposed on ourselves, it only remains to speak of the processes by means of which they obtain the sugar candy: but this manufacture, constituting, in France, at least, a part of the art of the confectioner, rather than that of the refiner, we shall only point out very summarily the labours by which they make it.

Sugar candy does not differ from sugar in the loaf except in this, that its crystallization instead of being produced by the stirring, must be effected by repose; and also, that it may be done more slowly, in order that the crystals be more regular, we have removed all causes of a too sudden cooling, and maintained the temperature of the place, where it is to a suitable degree for a time long enough. We have seen, on the contrary, that the operation known under the name of clouding in the manufacture of sugar in the loaf, has for its object to break the crystals, and to promote the cooling by renewing the surfaces. Thus they call regular crystallization, that by which they obtain sugar candy; and confused crystallization, that of loaf sugar.

The sirup having been clarified and filtered, is taken again into the reservoir of the clairée and carried into the cauldron, to be there baked to a suitable point. This is commonly, by the proof of the breath, weak or strong, according as we wish to obtain crystals larger or smaller.

We pour the baked sirup into copper basins, nearly hemispherical, the interior of which is perfectly polished. They are from fifteen to eighteen inches diameter at their edge, and six to eight inches deep. At about two inches below the edge, they are pierced on each side with eight or ten very small holes, through which a thread is passed, which goes from one edge to the other, passing through each of the holes. They stop these last either with paste, or by pasting paper on the outside of the basin, in order that the sirup, shall not flow through the holes.

The basins thus prepared are filled to an inch nearly above the threads, and carried immediately into a hot-house, the temperature of which is so high that the crystallization will not be complete till the end of six or seven days. After this time they remove the basins from the hot-house, and draw off the motherwater, that is, the sirup that remains liquid. They pour a little water in the basins, to wash the crystals which are spread over their bottoms. This water is put with the mother-water.

The bottom of the bed then presents a crystalline bed from six to nine lines thick. The threads which are covered with crystals have the form of garlands. They reverse the basins on a vase suitable to drain them well; after which they carry them again to the hot-house, that they may be well warmed. At the end of two or three days the sugar is dry; they take it from the hot-house, and remove it from the basins from which it is easily detached. It may then be put up for sale.

The mother-water enters into the manufacture of loaf sugar, like the bastards or lumps.

The tints more or less deep which many kinds of sugar candy exhibit, belong wholly to the purity of the sirup that has been used in making them. Sirup perfectly pure gives crystals entirely white.

Sometimes also they shade it in different manners by adding suitable coloring substances. It would lead us entirely from our subject to enter into the detail of these operations, which will be found, beside, in all works that treat of the art of the confectioner, into which they enter thoroughly.

Strawberry Ice Cream

from Eliza Leslie, *Directions for Cookery* (Philadelphia, 1837)

Take two quarts of ripe strawberries; hull them, and put them into a deep dish, strewing among them half a pound of powdered loaf-sugar. Cover them, and let them stand an hour or two. Then mash them through a sieve till you have pressed out all the juice, and stir into it half a pound more of powdered sugar, or enough to make it very sweet, and like a thick syrup. Then mix it by degrees with two quarts of rich cream, beating it in very hard. Put it into a freezer, and proceed as in the foregoing receipt. In two hours, remove it to a mould, or take it out and return it again to the freezer with fresh salt and ice, that it may be frozen a second time. In two hours more, it should be ready to turn out.

Lemonade

from *Enquire within upon Everything* (London, 1856)

Powdered sugar, four pounds; citric or tartaric acid, one ounce; essence of lemon, two drachms: mix well. Two or three teaspoonfuls make a very sweet and agreeable glass of extemporaneous lemonade.

Food for a Young Infant

from Sarah J. Hale, *Mrs Hale's New Cook Book: A Practical System for Private Families* (Philadelphia, 1857)

Take of fresh cow's milk one tablespoonful, and mix with two tablespoonfuls of hot water; sweeten with loaf sugar, as much as may be agreeable. This quantity is sufficient for once feeding a new-born infant; and the same quantity may be given every two or three hours, not oftener—till the mother's breast affords the natural nourishment.

To Make Bon-bons

from Angelina Maria Collins, *The Great Western Cook Book* (New York, 1857)

Have some little tin moulds, oil them neatly; take a

quantity of brown sugar syrup, in the state called a blow, which may be known by dipping the skimmer into it and blowing through the holes, when parts of light may be seen; add a few drops of lemon essence. If the bon-bons are prepared white, when the sugar is cooled a little, stir it round the pan till it grains and shines on the surface, then pour it in a funnel; fill the little moulds; when they are hard and cold, take them out and put them in papers. If you wish to have them colored, put on the coloring while hot.

Efferton Taffy

from [M. W. Ellsworth and F. B. Dickerson], *The Successful Housekeeper* (Harrisburg, PA, 1884)

This is a favorite English confection. To make it take three pounds of the best brown sugar and boil with one and one-half pints of water, until the candy hardens in cold water. Then add one-half pound of sweet-flavored, fresh butter, which will soften the candy. Boil a few minutes until it again hardens and pour it into trays. Flavor with lemon if desired.

Spinning Sugar

from Juliet Corson, *Miss Corson's Practical American Cookery* (New York, 1886)

Spun sugar is used to ornament large candied pieces of fruit and nuts, or nougat; for instance, the preceding piece, the chartreuse of oranges, might be covered with spun sugar after it is taken from the mould; or a pyramid formed of macaroons, cemented with white of egg; or any large ornamental combination piece built up of candied nuts, fruit, and macaroons; or such a stand of candy as is shown upon the table in the background of the accompanying engraving. The sirup is boiled to the degree called 'the crack,' and then a very little of it is poured from a spoon moved back and forth over an oiled knife held as shown in the engraving. The motion must be quick and steady; the spun sugar may be made in the long sections, or in shorter lengths; or it may be spun directly over the piece to be ornamented.

Ginger Pop

from Isabel Gordon Curtis, *The Good Housekeeping Woman's Home Cook Book* (Chicago, IL, 1909)

To two gallons of lukewarm water allow two pounds of white sugar, two lemons, one tablespoon of cream of tartar, a cup of yeast and two ounces of white ginger root, bruised and boiled in a little water to extract the strength. Pour the mixture into a stone jar and let stand in a warm place for twenty-four hours, then bottle. The next day it will be ready to 'pop.'

Select Bibliography

Abbott, Elizabeth, *Sugar: A Bittersweet History* (Toronto, 2008)

Abrahamson, E. M., and A. W. Pezet, *Body, Mind, and Sugar* (New York, 1951)

Appleton, Nancy, and G. N. Jacobs, *Suicide by Sugar: A Startling Look at Our #1 National Addiction* (Garden City Park, NY, 2009)

Aronson, Marc, and Marina Budhos, *Sugar Changed the World: A Story of Spice, Magic, Slavery, Freedom, and Science* (Boston, MA, 2010)

Aykroyd, W. R., *Sweet Malefactor: Sugar, Slavery and Human Society* (London, 1967)

Barnett-Rhodes, Amanda, 'Sugar Coated Ads and High Calorie Dreams: The Impact of Junk Food Ads on Brand Recognition of Preschool Children', Master's thesis, University of Vermont, 2002

Carr, David, *Candymaking in Canada: The History and Business of Canada's Confectionery Industry* (Toronto, 2003)

Chen, Joanne, *The Taste of Sweet: Our Complicated Love Affair with Our Favorite Treats* (New York, 2008)

De la Peña, Carolyn, *Empty Pleasures: The Story of Artificial*

Sweeteners from Saccharin to Splenda (Chapel Hill, NC, 2010)

Deerr, Noel, *The History of Sugar*, 2 vols (London, 1949)

Dibb, Sue, *A Spoonful of Sugar: Television Food Advertising Aimed at Children: An International Comparative Study* (London, 1996)

Duffy, William, *Sugar Blues* (New York, 1975)

Ebert, Christopher, *Between Empires: Brazilian Sugar in the Early Atlantic Economy, 1550–1630* (Leiden and Boston, MA, 2008)

Fraginals, Manuel Moreno, *The Sugar Mill: The Socioeconomic Complex of Sugar in Cuba, 1760–1860* (New York, 1976)

Galloway, J. H., *The Sugar Cane Industry: An Historical Geography from its Origins to 1914* (New York, 1989)

Gillespie, David, *Sweet Poison: Why Sugar is Making Us Fat* (Surry Hills, NSW, 2008)

Hollander, Gail M., *Raising Cane in the 'Glades: The Global Sugar Trade and the Transformation of Florida* (Chicago, IL, 2008)

Hopkins, Kate, *Sweet Tooth: The Bittersweet History of Candy* (New York, 2012)

Jacobson, Michael F., *Liquid Candy: How Soft Drinks are Harming Americans' Health* (Washington, DC, 2005)

Kawash, Samira, *Candy: A Century of Panic and Pleasure* (New York, 2013)

Keating, Giles, and Stefano Natella, *Sugar: Consumption at a Crossroads* (Zurich, 2013)

Kimmerle, Beth, *Candy: The Sweet History* (Portland, OR, 2003)

Krondl, Michael, *Sweet Invention: A History of Dessert* (Chicago, IL, 2011)

Lustig, Robert H., *Fat Chance: Beating the Odds Against Sugar, Processed Food, Obesity, and Disease* (New York, 2013)

Macinnis, Peter, *Bittersweet: The Story of Sugar* (Crows Nest, NSW, 2002)

Mason, Laura, *Sweets and Sweet Shops* (Haverfordwest, Pembrokeshire, 1999)

—, *Sugar-plums and Sherbet: The Prehistory of Sweets* (Totnes, 1998)

Mazumdar, Sucheta, *Sugar and Society in China: Peasants, Technology and the World Market* (Cambridge, MA, 1998)

Mintz, Sidney W., *Sweetness and Power: The Place of Sugar in Modern History* (New York, 1985)

Moreno Fraginals, Manuel, *El Ingenio* (Barcelona, 2001)

Moss, Michael, *Salt Sugar Fat: How the Food Giants Hooked Us* (New York, 2013)

O'Connell, Sanjida, *Sugar: The Grass that Changed the World* (London, 2004)

Osborn, Robert F., *Valiant Harvest: The Founding of the South African Sugar Industry, 1848–1926* (Durban, 1964)

Parke, Matthew, *The Sugar Barons: Family, Corruption, Empire, and War in the West Indies* (London, 2011)

Penfold, Steve, *The Donut: A Canadian History* (Toronto, 2008)

Richardson, Tim, *Sweets: A History of Candy* (New York, 2002)

Scarano, Francisco A., *Sugar and Slavery in Puerto Rico: The Plantation Economy of Ponce, 1800–1850* (Madison, WI, 1984)

Schwarz, Friedhelm, *Nestlé: The Secrets of Food, Trust, and Globalization*, trans. Maya Anyas (Toronto, 2002)

Schwartz, Stuart B., ed., *Tropical Babylons: Sugar and the Making of the Atlantic World, 1450–1680* (Chapel Hill, NC, 2004)

Siler, Julia Flynn, *Lost Kingdom: Hawaii's Last Queen, the Sugar Kings and America's First Imperial Adventure* (New York, 2012)

Strong, L.A.G., *The Story of Sugar* (London, 1954)

Warner, Deborah Jean, *Sweet Stuff: An American History of Sweeteners from Sugar to Sucralose* (Washington, DC, 2011)

Woloson, Wendy A., *Refined Tastes: Sugar, Confectionery and Consumption in Nineteenth-century America* (Baltimore, MD, 2002)

Yudkin, John, *Pure, White and Deadly* (New York, 2013)

—, *Sweet and Dangerous: The New Facts about the Sugar You Eat as a Cause of Heart Disease, Diabetes, and Other Killers* (New York, 1972)

Websites and Associations

Sugar Millers, Refiners, Associations, Organizations and Societies

American Crystal Sugar Company

www.crystalsugar.com

American Sugar Alliance

www.sugaralliance.org

Australian Sugar Milling Council

www.asmc.com.au

Brazilian Sugarcane Industry Association (UNICA)

http://english.unica.com.br

British Society of Sugar Technologists

www.sucrose.com/bsst

China Sugar Association (CSA)

www.csa.gov.cn/outline.asp

The International Society of Sugar Cane Technologists

www.issct.org

The International Sugar Organization

www.isosugar.org

South African Sugar Association (SASA)

www.sasa.org.za

The Sugar Association

www.sugar.org/about-us

Sugar Association of London and Refined Sugar Association

www.sugarassociation.co.uk

SugarCane.Org

www.sugarcane.org

Sugar Producer

www.sugarproducer.com

Tate & Lyle

www.tateandlyle.com

World Sugar Research Organisation

www.wsro.org

High Fructose Corn Syrup

Corn Refiners Association

www.corn.org

High Fructose Corn Syrup

www.sweetsurprise.com

Candy Companies

Cadbury

www.cadbury.co.uk

Hershey

www.hersheys.com

Mars

www.mars.com

Nestlé

www.nestle.com

Perfetti Van Melle

www.perfettivanmelle.com

Beverage Companies

Coca-Cola Company

www.coca-colacompany.com

PepsiCo

www.pepsico.com

Red Bull

www.redbull.com

Schweppes

www.schweppes.com

Sugar Research and Institutes

eRcane (Energie Reunionaise Cane)

www.ercane.re

Mauritius Sugar Industry Research Institute (MSIRI)

www.msiri.mu

Ponni Sugars (Erode)

www.ponnisugars.com

Sugar Processing Research Institute

www.spriinc.org

United States Department of Agriculture—

Sugarbeet Research Unit

www.ars.usda.gov

Vasantdada Sugar Institute

www.vsisugar.com

Photo Acknowledgements

The author and the publishers wish to express their thanks to the below sources of illustrative material and/or permission to reproduce it.

Alamy: p. 103 top (Cindy Hopkins); Bigstock: p. 2 (contents page) (luiz rocha); The British Library, London: pp. 021, 034, 044; ChildofMidnight: p. 114; Brandon Dilbeck: p. 090; Thomas Dohrendorf: p. 070; Courtesy of Kelly Fitzsimmons: pp. 3, 007, 008, 019, 020, 026, 028, 034, 036, 040, 041, 042, 043(top), 048, 054, 055, 085, 086, 099, 102, 109, 111, 131; Getty Images: p. 097; iStockphoto: p. 107 (JenD); Library of Congress, Washington, DC: pp. 015, 029, 043 bottom, 047, 051, 052, 072, 077, 130; Shutterstock: pp. 100, 101 (Roman Samokhin), 103 bottom (ValeStock); Stratford490: p. 088; Tup Wanders: p. 084; Wellcome Library, London: pp. 056, 127.

图书在版编目（CIP）数据

糖 / (美) 安德鲁 · F. 史密斯著；王艺蒨译 .
-- 北京：北京联合出版公司，2024.4
(食物小传)
ISBN 978-7-5596-7360-2

Ⅰ. ①糖… Ⅱ. ①安… ②王… Ⅲ. ①糖果－历史－世界－普及读物 Ⅳ. ① TS246.4-091

中国国家版本馆 CIP 数据核字（2024）第 023592 号

糖

作　　者：〔美国〕安德鲁 · F. 史密斯
译　　者：王艺蒨
出 品 人：赵红仕
责任编辑：李艳芬
产品经理：汤　成　甘　露
装帧设计：鹏飞艺术
封面插画：〔印度尼西亚〕亚尼 · 哈姆迪

北京联合出版公司出版
（北京市西城区德外大街 83 号楼 9 层　　100088）
北京天恒嘉业印刷有限公司印刷　　新华书店经销
字数 143 千字　889 毫米 ×1194 毫米　1/32　9.75 印张
2024 年 4 月第 1 版　　2024 年 4 月第 1 次印刷
ISBN 978-7-5596-7360-2
定价：59.80 元

北京市版权局著作权合同登记 图字：01-2022-7168号